3D Manufacturing Innovation

Hiroshi Toriya

3D Manufacturing Innovation

Revolutionary Change in Japanese
Manufacturing with Digital Data

Translated by Yukie Ito

 Springer

Hiroshi Toriya, PhD
President and CEO
Lattice Technology, Co., Ltd.
4F Hiei-Kudan Building,
3-8-11 Kudan-Minami, Chiyoda-ku
Tokyo 102-0074
Japan

ISBN 978-1-84996-709-9   e-ISBN 978-1-84800-038-4

DOI 10.1007/978-1-84800-038-4

British Library Cataloguing in Publication Data
Toriya, H. (Hiroshi), 1960-
  3D manufacturing innovation : revolutionary change in
  Japanese manufacturing with digital data
  1. CAD/CAM systems - Japan
  I. Title
  670.2'85

Previously published in Japanese by Nikkei Business Publications, as 3 次元ものづくり革新 (3D Manufacturing Innovation): デジタルデータが変える日本製造業 (Revolutionary Change in Japanese Manufacturing with Digital Data, by Hiroshi Toriya, 2006, ISBN 978-4-8222-1892-8.

Cover design: eStudio Calamar S.L., Girona, Spain

Printed on acid-free paper

9 8 7 6 5 4 3 2 1

springer.com

Foreword

Mr. Kentaro Kizaki
Nikkei Monozukuri Editor, Nikkei Business Publications, Inc.
Amidst the intensifying competition revolving beyond national borders in the manufacturing industry, what is the competitive edge required for manufacturers to survive? Competitive edge can be visible and invisible. "Visible competitive edge" means it is visible to customers, such as new mechanisms of products, materials, machining technologies, *etc.* It can also be called product innovation. In contrast, "invisible competitive edge" is competitive edge in the business process, in other words, competitive edge created by innovation of design and production processes, or by the establishment of mechanisms for manufacturing quality products quickly and inexpensively by the introduction of IT or human resource development.

Indispensable to this "invisible competitive edge" is no doubt the use of 3D for the design and manufacturing processes. With the growing use of 3D CAD in design departments, 3D data-based manufacturing is becoming more and more common. In design departments, 3D data is used as the material for verifying design, and in manufacturing departments, 3D data is used for machining and automatic assembly. 3D data is also used for enhancing the manufacturing process to facilitate the work of assembly operators. Procurement and marketing departments can also use 3D data for their procurement, sales, and logistics activities as predictors. In whichever case, 3D data is a tool which supports in ways invisible to customers of the manufacturing industry. The full and thorough use of 3D data will reinforce invisible competitive edge. According to Professor Takahiro Fujimoto of the University of Tokyo (Graduate School of Economics), "MONOZUKURI" (manufacturing or making in Japanese) is the transcription of design information onto media. He says, for example, an automobile is the transcription of design concepts of a vehicle onto a 0.8-mm-thick metal sheet. Important here is the fact that satisfying the customer is not the medium of metal sheet but the design information itself. And the means of conveying this design information from designers to production engineers are drawings or 3D data, *etc.*

Japanese Drawings: Designed to Facilitate Production Technology Plans

At the moment (2007), most 3D CAD software products used in design departments are from Europe or the USA. The use of 3D CAD enables designers to define 3D models precisely as well as convey shapes accurately to those applying production technologies. 3D CAD is very clear-cut, nothing is vague, so it tries to provide all the information required in the downstream process. This was why CAD was developed in Europe and the USA. On the other hand, the aim of drawings used in Japan has been for designers to relay design details to production engineers. The process of preparing these drawings consisted only of reconstructing design information so that production engineers can understand the information better. So the production engineers would look at the drawings, try to understand the intent of designers, and work on the production process. This creativity at the production side is what strengthens the foundations of the Japanese manufacturing industry.

In manufacturing, 3D CAD data is, needless to say, very useful. There are very keen efforts to realize "drawing-less" manufacturing by digitizing information transmission and abolishing drawings. However, it is risky to simply replace drawings with 3D CAD data. This is because the meanings of drawings and 3D data are totally different in production engineering.

If the strengths of the manufacturing industry to date are to be made use of, it is necessary to use 3D CAD data but, at the same time, apply a method which will reliably convey designer intent such as tolerance and important details to the production process. Such a reliable method would be lightweight 3D data as represented by XVL. Lightweight 3D data should not be taken as the simplified version of CAD data, because not only production engineering departments but also various departments can access this lightweight 3D data to learn about design intent. For this, large-scale assembly data must be viewable and easily accessed by virtually anyone.

Japan Leads the World in Use of 3D Data

Japan clearly leads the world in the use of 3D data. Lightweight 3D data XVL was developed by the Japanese company Lattice Technology (hereafter referred to as Lattice) and is growing more and more popular. Other lightweight 3D data software include Fujitsu's VPS and Digital Process' VridgeR. Though differing in the functions provided, Japan has pioneered the use of 3D data software. This can probably be attributed to the sophisticated skills of users of XVL and 3D tools, in other words, production engineering departments of Japanese companies.

It is natural for Japanese industry to reinforce its manufacturing strengths through IT. As European and American CAD software have already penetrated deeply into the Japanese market, there is not much value in developing Japanese CAD systems now. So in order to strengthen the Japanese manufacturing industry, we need software that matches Japan's manufacturing culture. One such software would be software enhancing collaboration between design departments and manufacturing departments such as production engineering. In order for Japan to carry out concurrent engineering the Japanese way, it will need mechanisms for production-related staff to participate in the design process from an early stage. This would be design review using 3D data as a tool for communication between different departments.

From the perspective of the partnership between design and manufacturing, design reviews embody the bottom-up approach where optimization proceeds by trial and error. In contrast, European and American software vendors propose the concept of realizing overall optimization all at once, called Product Lifecycle Management (PLM). PLM is a top-down approach where attempts are made to manage and use information in the product lifecycle from upstream to downstream to enhance the competitive edge. In reality, this approach for overnight reforms is sometimes difficult. When attempts are made to resolve a big problem, it is usually difficult to decide where to start from. It is therefore more realistic to start by accumulating CAD data and lightweight 3D data in the company's common database so that people requiring information can access it. Once more and more people are using the data, then it is time to enhance software and hardware. The fact that lightweight 3D data such as XVL allows such a bottom-up approach makes it advantageous in reinforcing Japan's strengths.

Preface

It is said that good users are essential to the birth of good software, because it is the discerning users who help foster software quality. This book discusses the uses of 3D data mainly in the Japanese manufacturing industry. Originally, 3D CAD, CAM, and CAE data was used exclusively for product design. However, in recent years, the Japanese manufacturing industry has used 3D data to revolutionize manufacturing processes. By using lightweight 3D formats such as XVL, Lattice Technology's eXtensible Virtual world description Language, Japanese manufacturers have improved production and laid the groundwork for innovative new methods of corporate communication.

This book discusses how leading Japanese manufacturers use 3D data in downstream processes, how the IT infrastructure required for this has been built, and some of the trial and error behind these developments. Each of the companies introduced as case studies are leaders in Japanese industry. It should be particularly interesting to European and American manufacturers to learn how their counterparts in Japan make use of IT to gain competitive strength. In fact, European and American manufacturers are starting to use 3D in downstream processes; this book includes examples from three leading manufacturers. It is interesting to note that the software described in this book, which supports manufacturing, a forte of Japan, was also developed in Japan, demonstrating that outstanding software is indeed nurtured by outstanding users.

We sometimes hear people talk about the CRIC cycle, which stands for crisis, response, improvement, and complacency. When faced with a crisis, people respond and try to fix the problem. The conditions improve, and then complacency sets in. But what happens if the solution is a "quick fix" that does not solve the underlying problem? People are complacent, but the risk remains – a trap we are all apt to fall into.

The CRIC cycle can sometimes be seen in the manufacturing industry. Manufacturers constantly strive to enhance quality, cut costs, and shorten delivery times. 3D CAD/CAM/CAE has been embraced as a solution to these challenges.

In the 1990s, Boeing started using 3D CAD to design its 777 family of aircraft. This endeavor, which involved intense collaboration with partner companies, demonstrated clearly the advantages of concurrent engineering using 3D design.

At the same time, China was seen to rapidly adopt 3D design, skipping the 2D CAD drawing step that other nations had gone through. Japanese industry experienced a sense of crisis and rushed to start using 3D CAD as well. This was how applications of 3D design in Japan started to shift into full swing. It is now expected that Japanese manufacturers will be able to innovate their production processes using the 3D CAD data that has accumulated in design departments. However, in reality, many companies do not seem to be fully utilizing the 3D CAD software which they have procured. In addition, many companies that have embraced 3D design are using 3D data only for checking simple 3D shapes and drawing illustrations very limited applications. It looks like these companies have fallen into the trap of the CRIC cycle. It is said that if the path from crisis to complacency is long, the path that follows is also long. This means that if the cycle prolongs, it becomes difficult to break away from the crisis. These companies therefore need to ask if they are content to just have installed CAD, or to just be using the 3D data for limited purposes, and if the improvement measures they have implemented are not simply quick fix solutions.

Adoption of 3D CAD incurs huge costs for procuring and installing expensive hardware and software, training costs for designers and engineers, and costs for changing business processes. However, often the 3D CAD data generated at such high costs is used only in design and manufacturing, which make up less than 10% of the whole IT domain. The other 90% sees no benefit from this data. Often this is blamed on the large size and complexity of 3D data which makes it difficult to use. However, things are changing with the emergence of lightweight 3D data formats and viewers in recent years, which is increasing the use of 3D data not only inside the company but also outside. This is a natural development because 3D data can be understood intuitively and is an optimum tool for communication. Even Microsoft Windows Vista is equipped with a 3D viewer function, which is expected to increase the visibility and importance of 3D data.

This book introduces methods of using 3D data to enhance competitive strength in manufacturing. Chapter 1 explains the current situation of 3D design in Japan, a source of competitive strength of the Japanese manufacturing industry. Chapter 2 describes the background of lightweight 3D data. Chapter 3 introduces the pioneering case study of SONY which describes how to build an information infrastructure for 3D data. Chapter 4 discusses the advantages of using general lightweight 3D data, and Chapters 5–13 are case studies of leading manufacturers that have innovated business processes using 3D data. The lessons learned from their efforts are summarized in Chapter 14, and the lightweight 3D tools that these companies used are explained in the two appendices.

This book hopes to capture the essence of using 3D by examining leading edge efforts in 3D data applications. Though 3D can be beneficial for limited applications, such an approach fails to capitalize on the benefits of 3D data. Only by standardizing 3D use across the enterprise can companies fully realize the value of 3D data and break the CRIC cycle.

The use of lightweight 3D data is an attempt to incorporate IT into manufacturing technologies. The goal of the use of 3D data is to eliminate all unnecessary work of designers and manufacturing staff so that they can concentrate on innovative work. In addition, by sharing knowledge from design and manufacturing with downstream departments, quality and productivity can be enhanced throughout the company. By taking readers through 3D data uses by pioneering companies, this book hopes to show how IT can be used to improve manufacturing not just in Japan, but all over the world.

Acknowledgments

In writing this book, I had the opportunity to speak with many users of XVL who provided valuable information and insight. My deepest thanks goes to Mr. Masashi Watanabe, Mr. Hiroshi Sekiya, Mr. Hideki Yoshii, and Mr. Taichi Tsukamoto of SONY Global Solutions, Mr. Junichi Harada and Mr. Shigeharu Ueyama of TOYOTA, Mr. Kiyotaka Yamamoto of NIKON, Mr. Hiroshi Takaya of YAMAGATA CASIO, Mr. Shigeki Yoshiwara and Mr. Nobuyoshi Mizuno of ALPINE PRECISION, Mr. Hisao Horibe of TOKAI RIKA, and Mr. Mitsuhiko Iwata and Mr. Hideo Kashiwakuma of CASIO. These are the users who have been able to make full use of and experience the advantages of 3D data through their tireless efforts to promote the use of 3D within their organizations. I would also like to express my gratitude to Mr. Larry Dietzler of L-3 COMMUNICATIONS, USA, Mr. Sebastien Jame of KVAL, USA, and Mr. Dieter Ziethen of MAN, Germany, for allowing me to introduce their leading-edge applications of XVL. I also thank my employees at Lattice Technology for helping check the Japanese draft of this book; Ms. Mayumi Matsuura, Mr. Kouji Yamato, Mr. Satoru Hatakoshi, Mr. Takeshi Yasuda, Mr. Koichi Kaneko, Ms. Ai Shibata, and Ms. Hitomi Saitoh. For the US and German case studies, I thank Mr. Shuji Mochida and Mr. Bill Barnes for their help and cooperation. In realizing the English version of this book, I am indebted to Mr. Junji Nagasaka, CEO of Toyota Communication Systems , Prof. Emi Miyachi of Cyber University, and Mr. Satoshi Ezawa, CEO of MetaLinc. I also thank Mr. Daichi Aoki for helping prepare this English version, to Ms. Yukie Ito for taking on the difficult task of translating the book, and to Mr. Marc Jablonski for English review based on his extensive knowledge of this industry. Finally, I thank the partner companies of Lattice, and all Lattice employees especially Mr. Tsuyoshi Harada, Mr. Kouichi Kobayashi, Mr. Yoshito Inoichi, and Mr. Masato Toho for their many helpful suggestions and firm support to realize this book.

Contents

1 Adoption of IT by Manufacturing Industry to Enhance Competitive Strength.................... 1
1.1 Tasks in Manufacturing and Ideal Uses of IT............................ 1
1.2 Current Situation of Use of IT in Manufacturing 2
1.3 Strategies to Secure Competitive Advantage and Use of 3D Data 5
1.4 Trends in Lightweight 3D Data Related Technologies................. 7

2 Trend Toward Use of Lightweight 3D Data 11
2.1 Designs Based on 3D CAD to Full Use of 3D Data..................... 11
2.2 Why Lightweight 3D Data, not CAD? 13
 2.2.1 Display of Very Large Data............................ 14
 2.2.2 3D Use in Documents 14
 2.2.3 3D Use in Drawings............................ 15
2.3 Use of Lightweight 3D Data Throughout the Company............. 16

3 SONY's Ideas on Expanding Lightweight 3D Data to Company-wide Use........................ 21
3.1 Use of 3D Data in Design and Manufacturing at SONY.............. 21
3.2 Introduction of Lightweight XVL 3D Data............................ 24
3.3 Construction of "3D Data Information Distribution Platform" 25
 3.3.1 Data Sharing........................ 27
 3.3.2 Data Distribution and Management 27
 3.3.3 Management of Original Drawing Data........................ 27
3.4 Business Process Restructuring Using Lightweight 3D Data........ 27
3.5 Future Plans 30

4 Benefits of Lightweight 3D Data........................ 33
4.1 Use of XVL in Design Review........................ 34
4.2 3D Parts Lists 38
4.3 3D Parts Catalogs 40

4.4   Animated 3D Visual Manuals 43
4.5   Sharing CAE Analysis Results 46
4.6   Sharing CAT Measurement Data... 47
4.7   Collaborative Design Using Lightweight 3D Data 49

5   **Design Review in Body Design: Case Study of TOYOTA MOTOR
     CORPORATION**... 51
5.1   Why is Design Review Necessary?.. 51
5.2   Design Review Using XVL.. 52
5.3   The Actual Design Review Process... 55
5.4   Applications and Development of Design Review 56
5.5   Advantages and Disadvantages of 3D Design............................. 57
5.6   Two Goals of Using XVL 58

6   **NIKON: Use of 3D Data as a Communication Pipeline** 59
6.1   Environment of Semiconductor Fabrication Devices................... 59
6.2   Design and Manufacturing Process Innovation with 3D Data....... 60
6.3   Difficulties Using 3D Data in Downstream Processes 63
6.4   XVL's Role as a Communication Pipeline.................................... 63
6.5   Security: A Pending Task.. 67

7   **YAMAGATA CASIO: Digital Engineering Practiced at Injection
     Mold Plant and Transfer of Technological Information** 69
7.1   Digital Equipment Market... 69
7.2   3D CAD/CAM and Network 70
7.3   Why 3D Design Alone is not Effective 72
7.4   Ideals of Design and Mold Fabrication 73
7.5   Introduction of Process Management System 74
7.6   Changes in Information Transfer Media with Increased Use
       of 3D Design 76
7.7   XVL-based Technical Information Distribution Key to Success .. 78

8   **ALPINE PRECISION: Report-less and Drawing-less
     in Mold Making**... 81
8.1   Weapons for Global Expansion and Delivery Time Reduction 81
8.2   Limitations of Business Activities Based on Drawings
       and Reports... 82
8.3   Use of 3D Data for Mold Design Review...................................... 83
8.4   Company-wide Sharing of Design Information............................. 84
8.5   Review by Mold Manufacturing Department............................... 87
8.6   Application of 3D Data to Manuals... 89

9   TOKAI RIKA: Visualization of Manufacturing Information Mold Making Using 3D Work Specifications 91
    9.1   Tasks and Solutions in Mold-making Departments at TOKAI RIKA 91
    9.2   Using XVL and Advantages 92
    9.3   Using XVL in the Manufacturing Department 95
    9.4   How 3D Has Improved Operations at TOKAI RIKA 96
    9.5   From 2D Drawings to 3D Drawings 97

10   CASIO: Creating Customer Manuals Using 3D Data 99
    10.1   After 3D Design Practice Started Kicking In 99
    10.2   e-Manual Project 101
    10.3   Driving Force Behind Use of 3D Data 104
    10.4   Online Data Reviews 105

11   KVAL: 3D Information Sharing and Its Effects at a Middle-scale Firm 109
    11.1   Use of 3D Data for Maintenance of Complicated Machines 109
    11.2   Opening the Door Between Design and Manufacturing 110
    11.3   Use of 3D Data Between Manufacturing and Technical Support.. 111
    11.4   Future Plans: Aiming at 100% 3D 113

12   MAN Nutzfahrzeuge AG: Promoting Company-wide Process Chain Using 3D Drawings 115
    12.1   Using 3D Data for Design, but 2D Drawings for Communication 115
    12.2   Aiming at 3D Communication Throughout the Whole Process Chain 116
    12.3   Selecting XVL for its Lightweight and Interactive Features 117
    12.4   Multi-use of XVL Centering Around Data Management Tools 118
        12.4.1   Internal Communication 119
        12.4.2   Communications with Suppliers 119
        12.4.3   Technical Illustrations 119
        12.4.4   Assembly Instructions 120
        12.4.5   Quality Assurance 120

13   Using 3D Data Successfully 121
    13.1   Best Practices for Successful Use of 3D Data 121
        13.1.1   Design Review (DR) with Lightweight 3D 122
        13.1.2   Eliminating 2D Drawings and Reports 122
        13.1.3   Communicating with Lightweight 3D 123
        13.1.4   3D Documentation 123
        13.1.5   Sharing of 3D Data on CAE and CAT Systems 124

13.2   Systems that Aid in Successful Use of 3D Data........................... 125
        13.2.1   System for Storing 3D Data... 125
        13.2.2   System for Ensuring Security... 126

Appendix A Development Ideology... 129

Appendix B Overview of XVL Products.. 137

Index 153

Chapter 1
Adoption of IT by Manufacturing Industry to Enhance Competitive Strength

The Japanese economy has finally started to gain back its strength after a prolonged recession which resulted in its so-called "ten lost years." From the lessons learned during the bubble economy, Japanese companies are now aiming at "lean," cost-efficient, and profitable businesses, and are developing a dislike for excess staff and facilities. On the other hand, the globalization of business is increasingly pressuring companies to replenish goods as soon as they have sold. To enhance brand value, it is also crucial for manufacturers to ensure high quality in their products. This has led to a situation where such goals as faster delivery time and quality enhancement need to be realized by a limited number of people in design and manufacturing. With the dwindling birthrate and aging population, there is a lack of trained labor. The training of employees with inadequate skills is a problem. People in design and manufacturing are starting to grow tired. It is in such times that companies/industry need to use Information Technology (IT), review the corporate business process, and have staff focus on creative work. The keywords of IT in manufacturing are process evolution by 3D design and use of 3D data. Successful use of 3D data throughout a company in the manufacturing industry has now become indispensable for building competitive strength.

1.1 Tasks in Manufacturing and Ideal Uses of IT

Most companies give, as their topmost priority, the creation of a high-speed development system to shorten the time from product planning to market release. This is a common task for all manufacturers, from digital home appliances which have short product lifecycle, to cars, their related parts, machines, and so on. Speed is the ultimate goal for companies engaged in the intense competition to survive. The aim of speedy development is to be first to market with products that are appealing to consumers. This requires quick discovery of new user needs and rapid production of appropriate products. A shorter development time often means lower development costs, which is a second reason why so many companies are working on speedy development. The introduction of 3D CAD in product development by manufacturing companies has therefore been largely motivated by the desire to increase development speed.

On the other hand, the adverse effects of speedy development are starting to stand out. The manufacturing site is too busy and personnel are exhausted. Few have enough time to think about next generation technologies. Also seen are growing problems such as lack of time for staff training and use of outsourcing which means that the accumulated knowledge does not remain inside the company. Recently, we read about the decline in manufacturing quality in Japan. It is at just such times that we need to build IT infrastructures and use IT to support those involved in design and manufacturing. Unfortunately, present 3D CAD is often found to make things much busier at the design site contrary to its initial aim. It is tough to actually make 3D CAD work in operations, and this process often imposes a burden on design engineers. It seems that despite the tremendous efforts to create 3D data, the resultant data can only be used for a narrow range of applications. In other words, the value of the data is not as great as the labor to create it.

At the same time, Japan is said to be a broadband internet superpower. According to the penetration rate of broadband internet per household by country at the end of 2005 as announced by France IDATE, Japan is fourth in the world, coming after Korea, The Netherlands, and Sweden. Japan owns some of the leading IT infrastructures in the world. Broadband internet networks mean high-speed access to the Internet.

There is, however, a serious problem. In March 2006, the World Economic Forum announced that Japan ranks 16th place in terms of response index to IT infrastructure. The response index to IT infrastructure indicates the degree of contribution by IT infrastructures to economic growth. This means that Japan is not putting its world-class IT infrastructure to full use. So the improvement of this response index is crucial.

Japan is in a very good position to increase its IT infrastructure index in the manufacturing industry, because Japan leads the world in the use of 3D data. In Japan, unique software that suits Japanese manufacturing practices and culture is being developed, and software which supports the Japanese style of manufacturing – basic design in Japan, production in Japanese plants, and release the products to the world – is readily available. The manufacturing industry in Japan can increase its competitive strength by effectively using 3D data on its world-leading IT infrastructure.

1.2 Current Situation of Use of IT in Manufacturing

Figure 1.1 shows how manufacturing companies invested in IT in 2004 and 2005. As you can see, the importance of 3D CAD/CAM/CAE is overwhelming. The importance of 3D design is well known and 3D design, manufacturing, and analysis has already been adopted by 70% of the companies. The next large growth can

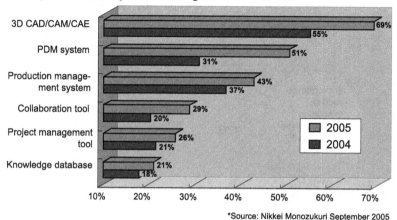

*Source: Nikkei Monozukuri September 2005

Figure 1.1 Manufacturing industry investment in IT infrastructure

be seen in Product Data Management (PDM), the database for the management of design data. The number of companies introducing PDM in their business grew from 40% in 2004 to above 50% in 2005.

This indicates that 3D design is becoming standard, and that the volume of 3D data is increasing. At the same time, systems are being constructed to ensure proper management of the accumulated data. Collaboration tools have a high growth rate. In 2005, nearly 30% of companies started to use these tools. The goal is to use accumulated 3D data for collaboration, marking the start of full-scale

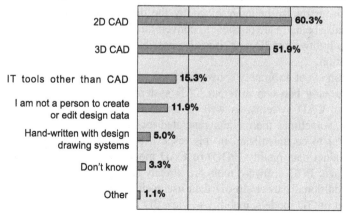

*Source: Nikkei Monozukuri Juli 2006

Figure 1.2 Increase in 3D CAD usage

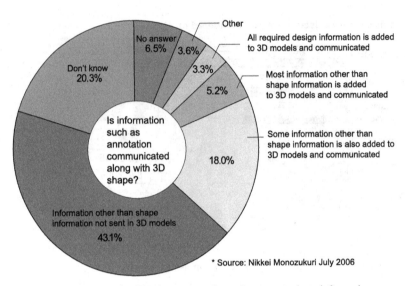

Figure 1.3   Drawings are still being commonly used to communicate information

efforts to share and reuse information. Figure 1.2 shows how design information is created. Use of 3D data is spreading in the same way as 2D CAD. However, as shown in Figure 1.3, 3D models are not used much as a means for information conveyance. Drawings are mainly used. To convey design information to the production process, 2D drawings are created from 3D models.

Figure 1.4 shows the reasons, namely 3D models are unable to include all the design information and information is sometimes not conveyed accurately. Not all design information to be conveyed can be incorporated in 3D models because they express only shapes. In order to convey design information, there is a need to convey product information which expresses the design intent, test information, *etc*. This requirement is gradually being resolved by the development of 3D CAD because it is gradually being made possible to convert dimensions and annotations defined by CAD to lightweight 3D data. It is also possible to use editing software to add this information.

Often information is not accurately conveyed because 3D model shapes cannot be exchanged accurately between different CAD systems. Because of accuracy differences between CAD systems, as well as the inconsistencies in the CAD modeling methods, sometimes the models generated include errors or the generated shapes turn out to be unrealistic. In this respect, there are growing efforts to enhance the product data quality (PDQ) of CAD models to reduce problems during data exchange. PDQ software tools are already available on the market. Figure 1.4 notes additional barriers to 3D data use such as the lack of development process based on 3D models, inability to view 3D models, *etc*. These problems stem from the nature of the design and production processes, and business reforms should be made top down to implement fully digital processes and

What are the problems when 3D models are used as means of communicating design information (multiple selections allowed)

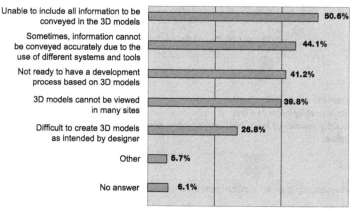

* Source: Nikkei Monozukuri July 2006

Figure 1.4 Barriers for using 3D models

collaboration between departments. Production activities using 3D data require reliable databases as well as mechanisms to distribute 3D data to the people who require it. Lightweight 3D data formats are progressing rapidly, gradually resolving problems faced on the system side. Business processes within the company must be adjusted to effectively use the accumulated 3D data to enhance work efficiency.

1.3 Strategies to Secure Competitive Advantage and Use of 3D Data

As Nicholas Carr talks about in his controversial book "IT Doesn't Matter," information technologies are maturing and can now be used by anybody. Computer technologies have also become standardized and prices continue to fall. How to effectively use IT is now a trivial issue, and IT no longer serves as a means for companies to differentiate themselves. Indeed the world has become a more convenient place now that software for email, Internet, word processing, and spreadsheets are available to all. And because they have become so common, such software no longer serves as the basis for corporate differentiation. CAD is also heading in this direction and becoming a commodity.

This brings us to the question of whether or not the effective use of accumulated 3D data can actually increase the competitive strength of the manufacturing industry. In this sector, advanced companies have been experimenting with light-

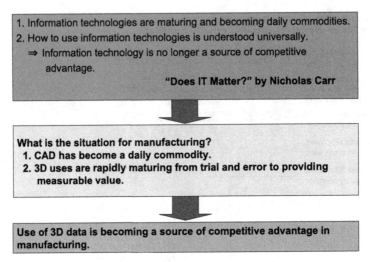

Figure 1.5   Information technologies becoming a commodity

weight 3D and have finally begun to see major results. Such companies have indeed started to build the foundations of competitive strength (Figure 1.5). As shown in Figure 1.6, strategies for securing corporate competitive advantage include cost performance strategies and differentiation strategies. 3D data can contribute to both. Cost performance strategies aim to provide products and services at lower costs than other companies, and one example is the continuous efforts toward improvement in manufacturing. Such efforts have paid off as more and more companies are seen to successfully enhance work efficiency by fully using 3D data. For example, illustrations in product manuals are automatically generated from 3D data, and these cost considerably less and are consistent in quality compared to hand-drawn ones. It is clear that use of 3D data contributes to cost advantage strategies.

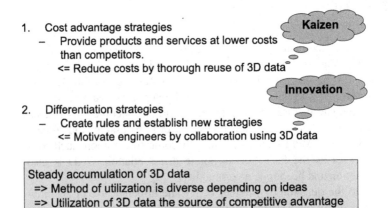

Figure 1.6   Corporate competitive strategies and use of 3D data

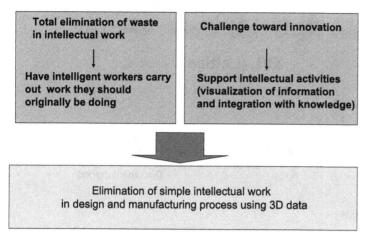

Figure 1.7 Acquire competitive edge in manufacturing

On the other hand, differentiation strategies aim to create new market rules and build up business with new strategies. This is called innovation. But such strategies are rarely born from a tired manufacturing site. Deploying IT to support collaboration between design and manufacturing increases the potential for innovation. The use of 3D data changes according to business type and activities. It can also be used in totally new ways. 3D data can also serve as the source of competitive strength of companies. Japanese manufacturing, "MONOZUKURI," is made up of both the "MONO" (product) and the "ZUKURI" (production). As shown in Figure 1.7, 3D data can contribute to both sides. For example, 3D data can reduce unnecessary work such as design and production preparations. There are many simple intellectual and manual tasks in design and manufacturing that can be eliminated by the use of lightweight 3D data. Employees who are freed from simple labor are able to work on intellectual innovation. Use of 3D data allows accumulation of feedback from the manufacturing site as visible information. This "visualization" of information promotes innovation.

1.4 Trends in Lightweight 3D Data Related Technologies

Let us take a look at the trends in technologies related to lightweight 3D data. Figure 1.8 shows the trends in lightweight 3D data formats. XVL technology, which was born in Japan, is used widely in manufacturing. Leading CAD vendors in Europe and the USA who have realized the usefulness of lightweight 3D data are also starting to move into this area. Dassault Systemes of France licensed XVL technology from Lattice for a lightweight 3D format 3D XML. Dassault Systemes' 3D CAD software, CATIA, is extensively used in the automotive and aerospace industries. At the same time, UGS of the USA is promoting a lightweight

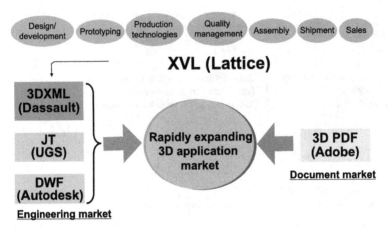

Figure 1.8   Lightweight data technology flow

3D format called JT. JT includes both polygon data for display and precision data for CAD data exchange. The largest PC CAD company, Autodesk, promotes a format called DWF for displaying lightweight 3D polygon data. Such CAD companies promote their respective formats using engineering-type applications.

In 2006, Adobe entered the market with 3D PDF (Portable Document Format). PDF is a very popular document format for distribution. 3D PDF adds Universal 3D (U3D) to PDF. U3D is a lightweight 3D format developed by Intel, Adobe, and the 3D Industry Forum, which includes Lattice. It is certified as a standard by ECMA, a European industry standardization organization. Support for U3D enables 3D PDF to display 3D models in PDF documents. The first U3D format was a polygon-based format, meaning that it used polygons (triangles) to represent geometry. This is a common technique for quickly displaying geometry on computers. The downside is that it takes a lot of polygons to accurately represent smooth 3D CAD surfaces, and that makes the file sizes of polygon-based formats very large. The goal for a lightweight 3D format is to have high accuracy with small file sizes, and companies are engaged in technological competition over this.

Interestingly, international standards are not always successful in the world of 3D. There have been many graphics standards like CORE, GKS, PHIGS, VRML, and X3D that have not taken off as anticipated. This may be due to the intense evolution of the technology; by the time standards are established, they quickly become obsolete.

Back to Figure 1.8 again. The battle around the standardization of 3D is intensifying. This battle differs from that around the standardization of videotapes, VHS versus Betamax, many years ago. In the world of hardware, tapes without compatibility cannot be played on a VCR. In the world of software, as long as the data can be converted from one format to the other, applications will be able to read it. Consequently, the essence of the problem of using lightweight 3D lies not just in the data format, but also in the applications for using the 3D data. The quality of the format affects the performance of the applications, but the applications must still meet end user needs.

Let us now discuss the efforts being made with XVL. XVL is already the most lightweight, accurate 3D format. XVL is now being expanded to express the design information required by manufacturing. The key users of lightweight 3D data formats are designers, who produce the 3D data, as well as downstream users who until now had no relation with 3D whatsoever. In general, for lightweight 3D formats to be useful they need to include not only 3D shapes but also non-geometric information required by the downstream users. Also, downstream applications need to be able to handle the 3D data and additional information.

Another important trend of the IT industry is the emergence of 64-bit PCs. These PCs are required by CAD users performing large-scale design and CAE users performing large-scale analyses. The 3-Gbyte virtual memory space on 32-bit PCs expands to 16-Tbytes on 64-bit PCs. In CAD and CAE applications where data is growing massively large, 32-bit PCs are no longer able to handle the burden. Already CAD manufacturers are starting to release products for 64-bit PCs on the market.

However current PCs are sufficient for ordinary users who just need to email, create documents, or make presentations. Powerful 64-bit PCs will only migrate downstream after hardware becomes inexpensive and most software can run in 64-bit environments. So what is going to happen until then? As Figure 1.9 shows, the design departments of leading manufacturers performing large-scale designs have 64-bit PCs, and CAD data is growing increasingly large. Downstream, it is necessary to display such massive data on 32-bit PCs. It will not be possible to distribute expensive 64-bit PCs and corresponding 3D CAD data to everyone involved, at least for the next few years.

Trends from 2007 to 2009

	Design process (CAD)	Downstream
Assembly manufacturer	64-bit PC	32-bit PC
Parts manufacturer	32-bit PC	32-bit PC
Requirements of lightweight 3D data	High precision (= surface data) Fast display speed	Lightweight (= small file size) Low memory use Fast display speed
Applications	Interference detection of huge 3D models	Free viewer 3D dynamic document Technical illustration

Need to easily process 3D data which is increasing in volume since 64-bit CAD introduction

Figure 1.9 Impacts of 64-bit PCs

Lightweight 3D which can express large volumes of 3D data lightly and accurately is growing in importance. Since most of the popular PCs are 32-bit systems, technology to display large volumes of 3D data on these 32-bit PCs is important. Consequently, lightweight formats like XVL have emerged that can display large volumes of 3D data on standard PCs. Over the past few years, the value of the existence of lightweight 3D data has increased more and more in companies.

The cost-performance of individual projects is said to depend heavily on the accumulation of IT assets and degree of use. In other words, companies with a solid IT infrastructure will reap greater profits from these projects than companies without such infrastructure. This is similar to the tool change process at the manufacturing site.

So the construction of IT infrastructure serves as preparation for the next process change. Preparations help produce results at high quality, short delivery, and low cost. Companies must prepare their IT infrastructure as standard IT strategies for business. 3D CAD and PDM are also gradually being established, and the use of lightweight 3D data should serve as the source of corporate competitive strength. In order to survive and win in the manufacturing competition, it is necessary to accumulate lightweight 3D data, share the information throughout the company and with related companies, and train key personnel to use it. All this points to the importance of building IT infrastructures based on the use of 3D data.

Chapter 2
Trend Toward Use of Lightweight 3D Data

The development of the network society has promoted the sharing of information in diverse areas of the manufacturing industry, changing the business processes of many companies. However, 3D CAD data, which forms the core of digital manufacturing, has not benefited much from network technologies. 3D CAD is characterized by massive data volume and complicated structures, and is also difficult to integrate with other associated data on the network. Though requiring huge costs to generate, until now CAD data has only been used in the CAD/CAM area, which comprises less than 10% of the whole IT industry.

The emergence of lightweight 3D data such as XVL is changing this situation. Industry is increasingly trying to change the overall business mechanism by converting CAD data to lightweight 3D data. This includes not only companies but also their partners and suppliers. So what is the significance of the existence of this lightweight 3D data and how should it ideally be used?

2.1 Designs Based on 3D CAD to Full Use of 3D Data

3D data is unique in that its simple shapes are easy for everyone to understand. In the case of design drawings, one has to be a design specialist to be able to conceive the actual shape. 3D display allows everyone to understand shape intuitively. This means that 3D data can serve as an excellent medium for communication. The Internet uses text, images, 2D animation, video, and audio to convey information clearly. 3D data will be next.

However, 3D data comes with a big problem, namely, the high costs needed to generate it. 3D CAD is very expensive. Compared to 2D data, 3D CAD data is difficult to handle because of the extra dimension of depth. 3D CAD also requires time and money to learn.

Still, 3D design has become the standard in many companies. In the manufacturing industry, 3D CAD is growing more and more popular and design departments are aggressively accumulating 3D data. It is only natural for companies to want to utilize this 3D data inside the company, with related companies, and for communication with consumers. Generally, a design department would have ten times its number of production and product-related partners, and these partners would have 100 times more consumers at their end.

If we can use 3D data for communication with product-related partners and consumers, *etc.*, we should be able to reform the business process of the manufacturing industry to a considerable extent. If we can effectively reuse the 3D data created in the design department, the downstream costs to generate 3D data will be zero. There are, however, two obstacles to enable everyone to use 3D data:

1. 3D CAD data is very large, and cannot easily be shared on the network.
2. 3D CAD systems to read the data are very expensive and complicated, and thus cannot be made available to everyone.

In order to overcome such problems, lightweight 3D data formats and software (viewers) to view them were developed. Generally, lightweight 3D viewers include free viewers for checking 3D shapes and products sold on the market for measurement and interference calculation.

HTML (Hyper Text Markup Language), the language for defining pages on the Internet is gradually being replaced by XML (eXtensible Markup Language), a more general language for representing information. XML can also define complicated 3D data structures and related design and manufacturing information.

By integrating the evolved 3D lightweight technology with XML, it should be possible to solve the above two problems. As shown in Figure 2.1, by converting 3D CAD data to lightweight 3D formats, general users who do not have CAD can use a viewer to access 3D data easily.

Lattice XVL uses XML to create a lightweight, accurate, web-friendly representation of 3D CAD data. This allows lightweight 3D data to be used across the "walls" of organization. XVL can compress 50 MB of CAD data down to 500 KB-small

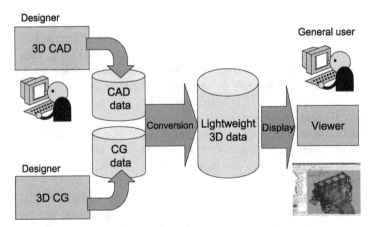

Figure 2.1 Use of CAD/CG data in downstream processes after conversion to lightweight 3D data

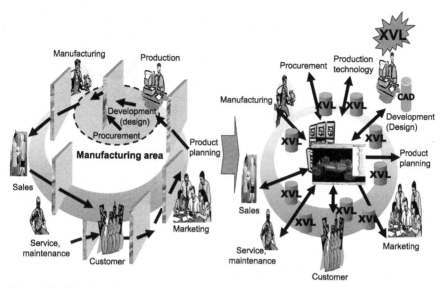

Figure 2.2 Transforming business processes using lightweight 3D data

enough share on the Internet. With a simple XVL viewer, anyone can easily use 3D data anywhere to carry out work. At Lattice, we call such environments allowing use of 3D easily "3D Everywhere."

3D data that has been released from the bonds of CAD can now be used beyond the design department as a clear communications tool (Figure 2.2). Manufacturing processes that involve diverse employees with different skills, employees of different companies, employees in different countries, and general consumers requires clear visual communication. 3D data released by the design department can provide clear visual communication and can improve business processes at the plant, in the maintenance department, in the marketing and sales department, and in other downstream departments.

2.2 Why Lightweight 3D Data, not CAD?

So why should we use lightweight 3D data instead of CAD data? In order to create lightweight 3D data, we need to convert it from 3D CAD data. Is it really necessary to go to the length of converting 3D data to generate lightweight 3D data? According to a pioneering XVL user, the advantages of lightweight 3D data are to display very large data, use 3D in documents, and use 3D in drawings (Figure 2.3).

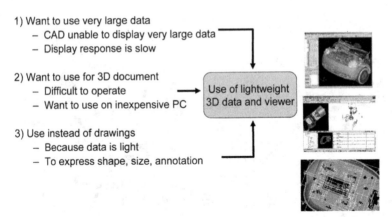

1) Want to use very large data
   − CAD unable to display very large data
   − Display response is slow

2) Want to use for 3D document
   − Difficult to operate   → Use of lightweight
   − Want to use on inexpensive PC   3D data and viewer

3) Use instead of drawings
   − Because data is light
   − To express shape, size, annotation

Figure 2.3 Why use lightweight 3D data?

2.2.1 Display of Very Large Data

In CAD, parts are designed by multiple designers, and the shape of the completed product is expressed by combining multiple parts. Depending on the design, data volume can be massive. In the automobile and airplane industries, which use high-end CAD, data volume is said to be 20 Gbytes for cars, and 5 Tbytes for airplanes. Unfortunately, CAD is unable to display such enormous data. When data reaches the level of several hundred megabytes, the display response of CAD drops markedly.

What about in terms of the number of parts making up a product? Generally, an industrial machine is made up of 3,000–5,000 parts, and a printer or photocopier about 5,000–8,000 parts. However, once the number of parts exceeds 5,000, the display response of CAD drops drastically. Of course, at such an enormous data size, real-time data sharing on the network becomes impossible. Lightweight 3D can solve this problem. For instance, XVL allows CAD data exceeding 10 Gbytes to be displayed.

2.2.2 3D Use in Documents

3D data is useful for documentation. By using illustrations generated from 3D data, technical documents can be created easily. Applications for creating illustrations easily from lightweight 3D data are improving rapidly. Also available now are software for creating and distributing 3D animations. However, keep in mind that the downstream users who display 3D animations are not CAD users. So inexpensive, easy-to-use viewers must be available for downstream users. The PCs of most of such users will probably be inexpensive, which are slower and have less memory than PCs for CAD. Lightweight 3D viewers must run with good performance on such PCs with low specifications.

2.2.3 3D Use in Drawings

There are also increasing efforts to use 3D data instead of drawings when conveying information from the design stage to post-processes. Most lightweight 3D data accurately expresses not only shape, but also the annotations, symbols, and dimensions used in design and manufacturing processes. Of course, since the 3D data is referenced by many users, it is necessary to convey the data via networks. This means that the data must be lightweight. In addition, an inexpensive viewer is all that is needed to check 3D shapes. Additional viewers are able to measure size and edit annotations.

Figure 2.4 compares 3D CAD and a viewer using lightweight 3D data. Departments which may need to create, edit, or change 3D data, such as the design department, have to perform 3D design on CAD. However, departments that only need to display 3D data other than design will benefit overwhelmingly from the use of lightweight data for the following three reasons:

1. No expensive PCs required
2. No sophisticated training required
3. Data can be exchanged via network

Downstream departments such as production technology departments, factories, quality assurance departments, service maintenance departments, and marketing departments just need to display lightweight 3D data. So, creating an environ-

	3D CAD	Lightweight 3D data+viewer
Shape generation	** 3D shape design	X No 3D shape design function ** Addition of mobile, configuration Information editing, and comments
Shape display	X Very large data cannot be handled * Difficult to import other CAD data	*** Very large data can be handled * Viewer data conversion and management required
Environment used in	X High performance PC required X Difficult to distribute through network	** Can use on inexpensive PC ** Easy to distribute through network
Implementation costs	X Expensive	** Inexpensive ** Free viewer
Education and training	* Special education and training are required	** Can use even with simple basic operation training

*** very good
** good
* good with condition
X not good

Can be used in production technology, plant, quality assurance, maintenance, marketing, sales

Figure 2.4 Departments suitable for using "lightweight 3D data + viewer"

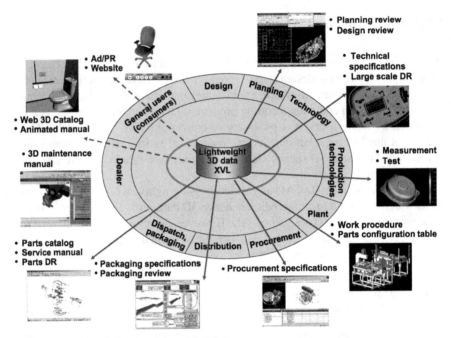

Figure 2.5   Areas of applying lightweight 3D data

ment that allows these departments to access and view 3D data is all that is needed to start using 3D data throughout the enterprise.

Figure 2.5 shows several potential applications for lightweight 3D data. Many companies carry out work restructuring based on 3D data. Planning departments use 3D for planning products, while design departments carry out design review of digital models that have been designed in 3D. Production technology departments use XVL data in measurement verification systems that analyze differences between CAD design data and actual work pieces. On factory shop floors, XVL is pasted into electronic reports, such as parts list and work specifications, thus sharply reducing work hours for creating the reports. Those in charge of viewing the reports also find that they understand the reports considerably more with the communication ability of 3D. 3D data can also be useful for creating parts catalogs in the service maintenance department and for web catalogs in the sales department.

2.3 Use of Lightweight 3D Data Throughout the Company

Once XVL starts to be used in various departments as illustrated above, many companies will face a data access problem. This refers to the fact that a general user may not know where a particular 3D data is stored. This will become a com-

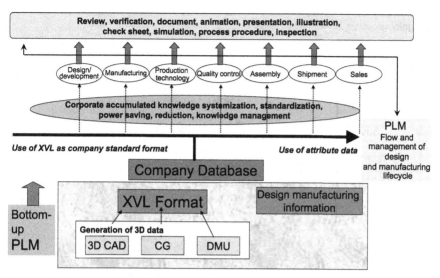

Figure 2.6 Overall optimization effects of XVL as corporate standard data

plicated problem to those outside the design department. If 3D data is to be used at the company level, the company must establish rules to indicate where to find the latest 3D data. One effective solution is to manage XVL within the company's standard database and enable it to be used by all departments (Figure 2.6).

Many companies have multiple 3D CAD systems. They convert all 3D data to XVL to allow people from various departments to refer to it. This in turn enhances the efficiency of a range of operations such as design review, documentation, illustration, sharing of design and manufacturing information, and so on by people from departments such as planning, manufacturing, production technology, quality control, and others. Data sharing throughout the company may raise security concerns, but security control mechanisms are already available for lightweight 3D data and will be discussed in a separate chapter.

Interestingly, lightweight 3D data has established a certain importance even within manufacturing departments, which are traditionally close to design departments. As shown in Figure 2.7, design departments use CAD for defining product shapes, and the shape data is saved as CAD models. CAD data replaces conventional drawings and is treated as official data (master data). This means that designs are saved as CAD data, and CAD models are revised when revisions are needed in the design. On the other hand, XVL serves as the master data for departments using 3D in the manufacturing process. There is a growing trend to write information such as manufacturing instructions and dimensions that are required in the manufacturing process in the XVL model, and this is maintained as master data. Through such trends, manufacturing processes are benefiting from 3D data.

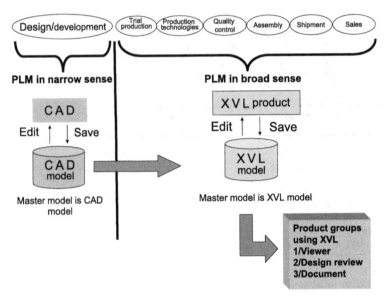

Figure 2.7 Applications of XVL

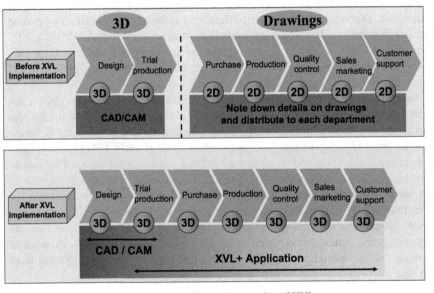

Figure 2.8 Changes in work before and after implementation of XVL

In the past, this was done using drawings. Information such as manufacturing instructions and attribute information required for manufacturing were individually written by different departments on the paper drawings from design. This means that manufacturing-accumulated knowledge was scattered over countless paper drawings. Replacing this with lightweight 3D data enables manufacturing data to be digitally integrated with the 3D data. This provides benefits such as accumulation of manufacturing knowledge in XVL, feedback of manufacturing information to the design department, and central management of information scattered over different drawings (Figure 2.8).

Furthermore, it provides a mechanism to allow access to this data by all who require it. As a result, the data can be used for design review, for design verifications, and for documentation. This means that efficiency can be enhanced by IT and the benefits of 3D design enjoyed even in departments downstream from manufacturing.

Chapter 3
SONY's Ideas on Expanding
Lightweight 3D Data to Company-wide Use

Lightweight 3D data is useful throughout design and manufacturing. Of course it also benefits local applications. However, it is always important consider overall optimization when promoting 3D data.

SONY is a company that has practiced this. They started using XVL in 2003 and learned how to use 3D data in their trial production processes. They learned that it was not easy to achieve best practice; people in downstream processes found it difficult to obtain 3D CAD data during the design process as well as identify which is the correct 3D data. To resolve these problems, SONY built a 3D data information distribution system to exchange XVL data with all their employees. Because they had such an information infrastructure, any department could use 3D data freely when they wanted. The company also launched projects to promote use of 3D data at the same time, which helped increase awareness of the usefulness of 3D data amongst their employees.

Such endeavors at SONY were pioneered by SONY GLOBAL SOLUTIONS (SGS), which provides the IT infrastructure for the whole of the SONY group. Leading these endeavors were Mr. Watanabe, head of the Engineering Solution Division, and Mr. Sekiya, head of the Engineering Information Solutions Division. The person responsible for building the actual infrastructure was Mr. Yoshii. Mr. Tsukamoto supervised the introduction and promotion of XVL. This chapter discusses how the leading Japanese company SONY started to use 3D data. Their experience in building a 3D data distribution infrastructure should serve as an important reference to companies planning company-wide use of 3D data.

3.1 Use of 3D Data in Design and Manufacturing at SONY

The key manufacturing businesses at SONY include TV, VCR, camcorder, digital camera, PC, *etc.* This leading company designs, manufacturers, and sells a diverse range of digital equipment. These high-tech devices are designed using mechanical CAD, electrical CAD, PDM/CAE, and systems to support the development of em-

21

bedded software. Mr. Watanabe's Engineering Solution Division and Mr. Sekiya's Engineering Information Solutions Division support the use of IT for the design of digital equipment at the SONY group. These divisions promote the restructuring of the design process of machines, electrical, and embedded software from two aspects: the use of state-of-the-art design technology and the construction of information infrastructure serving as the foundations for its use.

SONY starting using CAD from the 1980s for the design of digital equipment. Their management style of engineering tools has been transforming from "central to distributed and back to central" along with IT technological trends. The 1980s was an era of central management of host bases, and the 1990s an era of distribution by downsizing to PCs. Though SONY succeeded in realizing low costs by downsizing, distribution resulted in a tradeoff both technically and systematically from the viewpoint of Total Cost of Ownership (TCO). Then in the latter half of the 1990s, to realize global optimization in the network environment, they started to change back to central management, and this still continues to today. In the latter half of the 1990s, the company implemented PC-based 3D CAD, and began to carry out extensive in-house CAE analyses. As a result, designers themselves gradually started to encourage digital manufacturing using 3D data (Figure 3.1).

Figure 3.2 shows the concept of PLM from the viewpoint of mechanical CAD at SONY. The aim of the PLM system is to maximize the product-release speed and to design products unique to SONY. The PLM system requires mechanisms that allow access to product-related information. It is important to note that this design platform can be shared and used throughout the company, instead of being a CAD tool. It therefore serves as a design system incorporating design know-how instead of merely being a shape-creation tool. For this reason, ideally, 3D data allowing

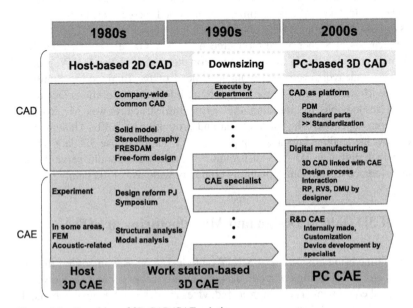

Figure 3.1 Transition of 3D CAD/CAE solution

Viewpoints from Engineering on Realizing PLM

| Shorten TTM | + | SONY way of product design |

Input horizontal distributed elements into product design process

Construct environment for accessing centralized
product-related information

> ➢ Transit engineering system environment to SONY's standard design
> platform
> ➢ Transit from CAD as tool for shape definition to digital design system
> with design knowledge accumulation
> ➢ Transform from 3D CAD as tool of mechanical designer to integrated
> CAD, which can simulate product design itself
> ➢ Design information system supporting uniformity, concurrent,
> horizontal distributed functions

TTM=Time to market

Figure 3.2   Core of PLM supporting SONY's product design

simulation of product design itself should be used and an infrastructure ensuring
the consistency of 3D data built. This example demonstrates that the establishment
of a design information infrastructure helps promote optimization by horizontal
distribution of design and manufacturing functions not only within the group,
SONY in this case, but also with partner companies.

Figure 3.3 shows the ideal use of 3D data as exemplified by SONY. Work can be
carried out while referring to 3D data during planning, design, prototype, mass production, service and support – even out to mold subcontractors. As a result, all CAD
data is converted to lightweight XVL and distributed through the company-wide
information distribution platform. Even CAD data that is in the design process and

Targeted goal

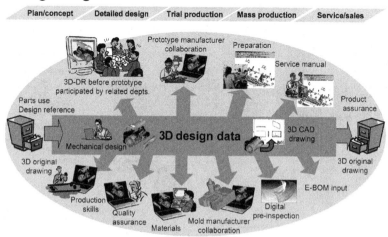

Figure 3.3   Best practice of lightweight 3D data

not yet approved is converted to XVL and made available to those who need it. However, before SONY was able to reach this stage, the company repeatedly evaluated XVL and verified the benefits of its use in actual work. Let us look at the company's decision to use XVL.

3.2 Introduction of Lightweight XVL 3D Data

Figure 3.4 outlines the flow of design data before SONY started to use XVL. Although the company was gradually incorporating 3D into their design work, they still used drawings after the prototyping phase. Once a product was completed, they would compare it with the drawings to confirm that production was carried out correctly. Work therefore focused on the downstream process.

They could identify problems in the downstream process, but it was important for them to identify these problems in the upstream process to improve the design work. In 2002, SONY carried out a technical evaluation of lightweight 3D data and decided to adopt XVL because of its high accuracy, light weight, web distribution function, low cost, and reliability amongst users. From 2003, they started services to convert from CAD to XVL. Figure 3.5 shows the mechanism of automatic processing conversion actually constructed at the company. Using this mechanism, they conducted several prototyping projects, and verified that XVL can contribute to improving operations.

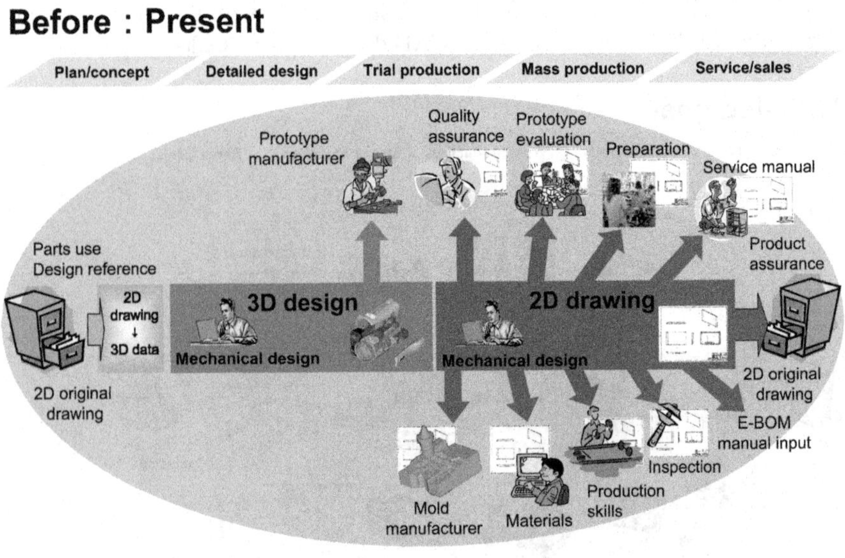

Before : Present

Figure 3.4 Initial design and manufacturing data flow

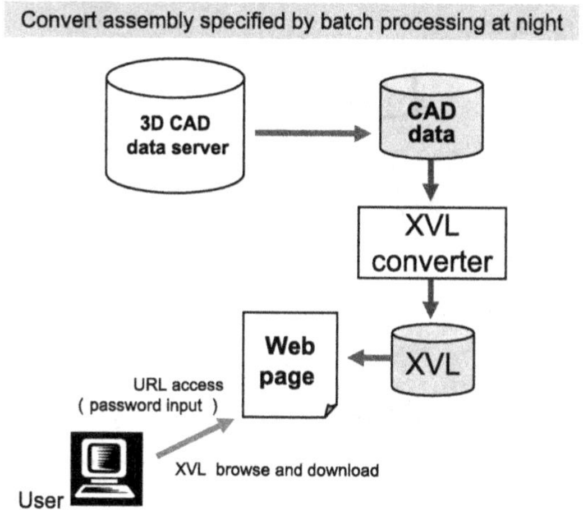

Convert assembly specified by batch processing at night

Figure 3.5 Mechanism of automatic conversion from CAD to XVL

However, the scope of applications of XVL did not broaden readily. It was not easy to reach the ideals shown in Figure 3.3 as standard design and manufacturing processes. There were two reasons for this. The first was that the XVL conversion mechanism was overburdened due to the large number of conversion requests. The other was the users of the 3D data could not really feel the benefits of using 3D data, so they were not very motivated to use 3D data.

In order to resolve this problem, SONY decided to construct a company-wide "3D data information distribution platform." At the same time, it launched a project to discuss the ideals of effective business processes using 3D data.

3.3 Construction of "3D Data Information Distribution Platform"

They faced obstacles when they actually tried to use 3D data because not all designers were in the state to disclose 3D data at all times. In some cases, the 3D model was not changed when the design changed. There existed three major problems as seen from the downstream process perspective (Figure 3.6):

1. Unable to obtain XVL easily when required
2. Unable to find latest 3D data for tasks such as preparation of manuals
3. Unable to determine which 3D model is the approved one

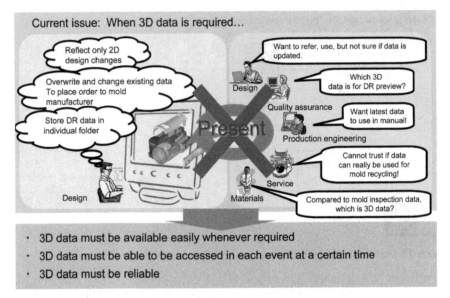

Figure 3.6 Problems in use of 3D data in post-processes

SONY therefore decided to construct a company-wide mechanism which allowed its users to access 3D data. This was the "3D data information distribution platform" shown in Figure 3.7. XVL was mounted on this platform to distribute 3D data. The platform is outlined below.

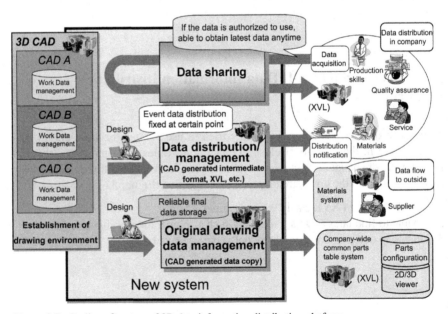

Figure 3.7 Outline of system of 3D data information distribution platform

3.3.1 Data Sharing

Those authorized to access data are able to obtain XVL anytime they want. These authorized persons are chosen according to the product. Only those related to the manufacturing and quality assurance of the product are allowed to access the 3D data. If there is an approved request to reference the data, the system automatically converts the 3D CAD data to XVL. As a result, people are even able to acquire data in the midst of design. This allows designers to verify their work and those of others from an early stage, and the production technological department is also able to start preparations from an early stage.

3.3.2 Data Distribution and Management

Notifications of appropriate 3D data distribution are sent to those who require it for uses such as prototyping, manual preparation, *etc.* This enables those requiring 3D data to access the required data. The data is available as CAD, XVL, and general 3D formats such as IGES. Lightweight XVL is frequently used in downstream processes. Manuals are prepared at the same time as design. Before, those requiring data had to ask individually and convert to XVL. Now the 3D data distribution has considerably reduced unnecessary work and distribution errors.

3.3.3 Management of Original Drawing Data

The design-approved final data is stored in and made available by the 3D data distribution system – drawings in PDF and 3D data in XVL. These two types of data are in formats that allow referencing from company-wide common parts lists. Verifications to check legitimacy can be performed using this final data. From a downstream perspective it is always clear what lightweight XVL data is final.

In this way, those requiring data can browse the latest design data by obtaining approval to access the design side. As long as the 3D data is sharable, XVL can be accessed from any manufacturing site in the world.

3.4 Business Process Restructuring Using Lightweight 3D Data

At SONY, efforts have continued since 2003 to make full use of 3D data in business operations. Reviews are performed by attaching animations to XVL to facilitate understanding of product functions and structure. XVL is also used for discussing designs with related departments such as manufacturing and service. Such activities gradually revealed the merits of XVL, such as easy to understand structure and functions.

With a better understanding of shapes during design meetings, the staff at SONY started to discuss improvement details more enthusiastically, and were even able to reduce downstream process steps. SONY also tested the use of XVL for preparing manufacturing instructions. In some cases, they found that the time taken to draw the actual object could be reduced from one week to one day.

The company's next task was to expand the use of XVL throughout the organization. SONY therefore launched a project leverage 3D throughout the company. In 2004 the design working group started meetings to discuss how to best utilize 3D data. Since this working group touches all SONY products, it includes representatives from most departments. This allowed success cases in specific product fields to be expanded to other product fields. The use of a common Bill Of Materials (BOM) system throughout the company is a strength of SONY, and this standardized system allows successes in one department to be repeated in other departments.

Whether 3D data can be utilized in a certain company activity was verified one by one by setting down target areas for XVL use (Figure 3.8). Usually, the introduction of new business processes depends on personal motivation. Unless a staff member is keen on the new process, the actual implementation in the company will be hard. In addition, if the users do not see the benefits then new methods will not be adopted. At SONY, the activities of the SONY working group therefore served as the motivating force for this establishment.

One of the missions of SGS is to construct a mechanism which facilitates the acceptance of new technologies at the site of manufacturing when new technology is introduced by the use of IT. For example, in reviews of design using 3D data, it is difficult to always have experienced key persons participate from the manufacturing department, and, at times, young engineers participate for the purpose of on-the-job training. Of course, it may not be an easy task to discuss 3D utilization in depth with veteran designers.

However, XVL allows data to be verified beforehand at the site as well as studied before meeting with the design staff. Such activities were supported in the SONY working group. Gradually these design review meetings became more and more active, and established new methods of using 3D data within SONY. With the launch of a new 3D information distribution mechanism, SONY was now able to extend the uses of 3D data to its global bases. In the past, it was difficult to request XVL conversion (from CAD data) if an engineer of a manufacturing division did not know the designer. Now, the new mechanism allows anyone to obtain XVL.So, even if the manufacturing staff do not personally know the designer, and even if there are language barriers or geographical distance, they are able to obtain and display 3D data freely. Also through telephone conferencing, engineers from anywhere in the world can hold design review meetings.

Two examples of the advantages of using 3D discussed during the SONY working group meetings are prototype review (Figure 3.9) and work standards. During prototype reviews, product structures are increasingly studied by 3D before the review meetings. People who are unable to visualize the shape just from

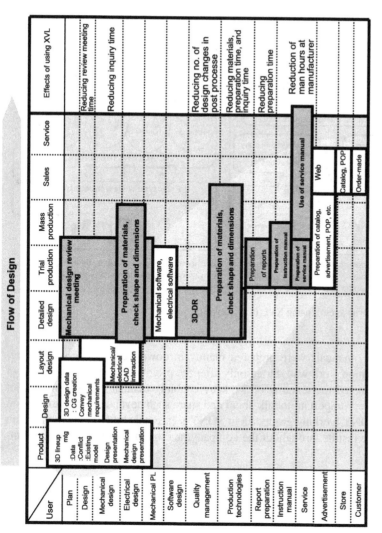

Figure 3.8 XVL utilization zone map

drawings, are more motivated to participate in meeting discussions, and as a result problems are being discovered and resolved at an early stage. The result is better design quality and shorter development times.

Apparently at SONY, when preparing work standards in the past they would go and borrow prototypes, take pictures of these prototypes with a digital camera, and use the image data to prepare drawings. However, pictured image data cannot be edited (such as deletion of unnecessary portions, changing of object direction, etc.). Now, use of XVL facilitates data editing as well as allows reuse of data even if the assembly procedure has been changed. In one case, SONY was able to cut drawing time by 30%.

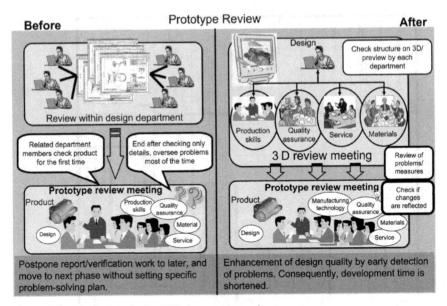

Figure 3.9   Advantages of using XVL in prototype review

Previously, manuals were prepared mainly in downstream processes and detailed preparation work was started only after the prototype was ready. The use of 3D data promoted the leveling of work by enabling manual preparation to start earlier. It is very important to talk about and share such advantages of introduction within the SONY working group. As in SONY's case, this built up the momentum to expand the results throughout the company, and build a platform for distributing 3D data.

3.5 Future Plans

In this way, SGS has been able to construct their 3D data information distribution platform, restructure their design business processes throughout the SONY Group, and continue to produce best practice scenarios on the use of 3D. The 3D infrastructure enables people requiring information to access 3D data.

As a result, departments throughout the company are producing significant achievements by implementing improvements. Spontaneous improvement efforts by staff, such as converting past models into 3D data so that the manufacturing department can refer to the data anytime, are increasingly being seen. Also gradually in downstream processes the "use of 3D data is taken for granted." In future developments, SONY feels they will need to add new information to the lightweight 3D data. This includes parts numbers, model names, and other product specifications that will help them make more effective use of the information.

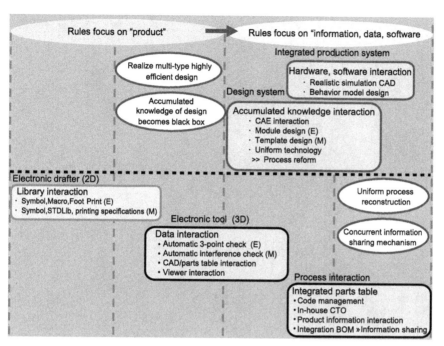

Figure 3.10 Advancement of CAD use

SONY says that, as a future goal, they hope to build digital models that allow simulation of the behavior of products and, at the same time, perform transforms to switch from rules centering around objects to rules centering on information, data, and software (Figure 3.10). They think it is ideal for 3D models to operate like the actual product according to user instructions to enable verification of functional
ity and convenience of use. Once digital models allowing uniform treatment of information on product mechanism and software to drive the mechanism are ready, they say they should be able to verify operations on the PC, as well as use it for early-stage marketing before release.

With their 3D data distribution infrastructure and experience in data use, SONY is steadily working on their next challenge.

Chapter 4
Benefits of Lightweight 3D Data

According to the US Department of Commerce, the US GDP grew by 4–5% between 1996 and 2000, of which 1.4% is said to have been contributed by IT. There is also statistical data which indicates that the US manufacturing industry has multiplied their productivity per person by more than ten times over the last ten years. In contrast, Japan's GDP growth rate was a mere 1% during its "ten lost years". The USA's innovative use of IT is well known, while Japanese industry needs to apply IT more aggressively to grow. Against this background, the Japanese manufacturing industry has started to apply lightweight 3D data to enhance business efficiency, and manufacturers of cars, electrical equipment, precision machinery, and machines are gradually beginning to enjoy the advantages of lightweight 3D data in routine business activities such as preparing work instructions, sharing information between designers and engineers, verifying design concepts, and sharing information with suppliers and customers (Table 4.1). This chapter discusses the benefits of using lightweight 3D data in various areas of industry. This chapter also describes conventional manufacturing methods and compares them to manufacturing methods using lightweight 3D data such as XVL.

Table 4.1 Examples of user applications of XVL

Company	Outline of user applications
NIKON	Uses XVL for design review (DR) of very large 3D models in the car body design process. The company has cut costs by using the automatic interference check function to discover and resolve problems at an early stage. They are also enhancing design quality by thoroughly reviewing problems during DR meetings.
TOYOTA	Uses 3D CAD to design semiconductor manufacturing machines containing hundreds of thousands of parts. The company is successfully increasing work efficiency by holding product development meetings and preparing manuals using XVL. Currently, XVL is the cornerstone of communication at the company.

Table 4.1 (continued)

Company	Outline of user applications
YAMAGATA CASIO	Has increased production efficiency by 30% by adding XVL to manufacturing and assembly specifications. The company also added XVL to the process management system, improving the process management of die and mold making. By incorporating information such as tolerance and finished state in XVL, the company has successfully established a 3D drawing-less culture.
ALPINE PRECISION	By replacing reports and drawings prepared during die and mold making with XVL, the company has enhanced information sharing between the designs and manufacturing departments. With XVL they are able to prepare for manufacture from an early stage, enhance die and mold quality, and enhance design quality by feeding back manufacturing information. The company is also using XVL for information exchange with overseas bases.
TOKAI RIKA	Has increased manufacturing efficiency by "visualizing" die production information with XVL. At the company, experienced engineers draw up specifications using 3D data and manufacturing is carried out according to these. By distributing XVL data, they have enabled production, inspection, and manufacturing departments that do not have CAD systems to provide design feedback.
CASIO	Has a system for compiling instruction manuals and parts catalogs using 3D data. This company uses XVL for most products. By developing a system which allows anybody requiring 3D data to convert to XVL via the web when they want it, they are smoothly using 3D data in downstream processes.
L-3 COMMUNICATIONS (USA)	By standardizing use of 3D data amongst all suppliers and partner companies, this company has sharply reduced costs. They also provide their customers with 3D process animations for product assembly. Customers can use these animations for product maintenance.
KVAL (USA)	Have dramatically increased maintenance efficiency by managing customized product information using XVL. This company has also shortened the technical training of new engineers by effective use of 3D data.
MAN (Germany)	Have incorporated XVL into their company-wide product data management system, dramatically improving communication. Also installs XVL Player in a large number of PCs at the company, realizing 3D information transfer.

4.1 Use of XVL in Design Review

Professor Takahiro Fujimoto of the University of Tokyo says that there are two types of manufacturing industries: combination (modular) manufacturing and collaborative (integral) manufacturing (Figure 4.1). With the combination manufacturing, the final product is made by assembling all the pre-designed parts. Each part serves a discrete function, so a product can be made just by combining the parts. One example of a combination manufacturing product is the PC. On the other hand, in collabo-

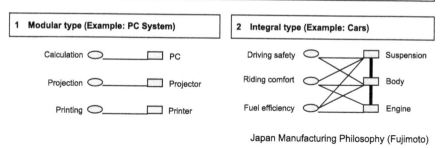

1. Combination type (modular)
- Produce individual products that are optimized for single functions and can be combined in diverse ways to produce optimal systems.

2. Collaborative type (integral)
- Produce integrated products that require simultaneous adjustment and optimization of individual parts to achieve optimal performance.

Figure 4.1   Types of manufacturing industry

rative manufacturing the component parts of a product must be adjusted and optimized together in order to achieve the best performance. The Japanese automotive industry is particularly adept at collaborative manufacturing. Since collaborative manufacturing products require integration and optimization of diverse parts, communication between the designers and engineers involved is very important. Therefore, Design Review (DR) is a critical process for collaborative manufacturers. Without DR, locally focused designers would not understand the overall system – a major cause of design errors. Design errors in turn result in unnecessary and costly design changes. In particular, design changes after production begins incur massive costs and time delays.

During DRs, everyone involved in the whole manufacturing process from design to production technologies, mass production, maintenance, *etc.* gather to discuss the details using 3D data. They look for design errors, determine if the assembly process would be smooth, and investigate possible manufacturing problems. The results are fed back to the design department to increase the design quality in the design stage (Figure 4.2). This means that problems that may occur in the downstream processes are predicted and prevented. Use of lightweight 3D data dramatically increases the efficiency of DR itself. Not only does it allow massive volumes of 3D data to be handled easily during DRs, but it also enables data sharing over different networks. More importantly, the use of 3D data enables DR to be made mandatory, so that design quality problems can be thoroughly detected and resolved in the design processes. This minimizes the number of problems that are discovered in downstream processes.

At collaborative manufacturing companies, data volume becomes massive when products are expressed in 3D CAD, making it incredibly difficult to review the data during DRs. The heavier the 3D data, the slower the viewer response. Current models are so large that it is impossible to display all the data in a single CAD session. Therefore conventional DRs are carried out using 2D drawings and cross-sections

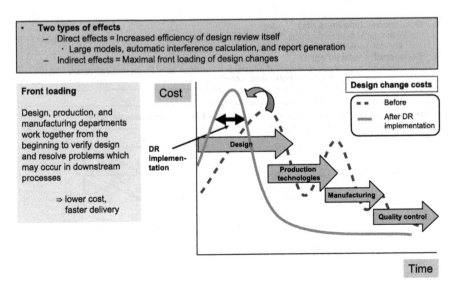

Figure 4.2   Effects of design review

that are prepared beforehand by the designers. Such DRs not only take a long time to prepare for, but also cannot be participated in by designers or engineers who are not drawing-literate (Figure 4.3).

Figure 4.4 shows the benefits of DRs using XVL. In the past, when part interference used to be checked by CAD, it was necessary to remove unimportant parts from the CAD data in order to be able to generate the 2D images and cross-sections. It was also necessary to visually check the cross-sections to find interference problems. This results in problems such as overlooked interferences and extra

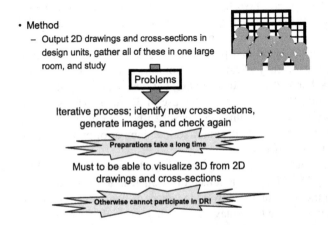

Figure 4.3   Conventional design review methods

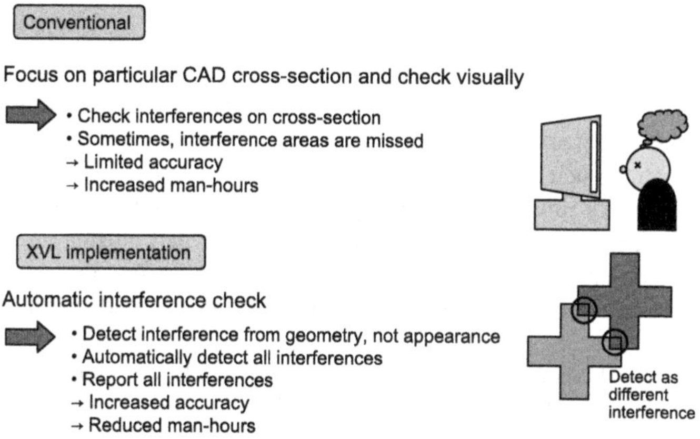

Conventional

Focus on particular CAD cross-section and check visually

➡ • Check interferences on cross-section
  • Sometimes, interference areas are missed
  → Limited accuracy
  → Increased man-hours

XVL implementation

Automatic interference check

➡ • Detect interference from geometry, not appearance
  • Automatically detect all interferences
  • Report all interferences
  → Increased accuracy
  → Reduced man-hours

Detect as
different
interference

Figure 4.4 Benefits of design review with XVL studio (1)

man-hours. XVL resolves all of these problems by automatically detecting all interferences on the complete model.

XVL can also perform clearance checks. Designers generally know from experience what clearance values are necessary to prevent defects. XVL can automatically detect and report locations that are below these values. As shown in Figure 4.5, XVL is able to calculate interference in large models without having to subdivide the data each time. XVL can also calculate interference by offsetting surface data. This is useful for checking car body surface data. XVL therefore reduces the time and increases the accuracy of interference and clearance checks. By pinpointing problems in the design stage, XVL makes it possible to resolve them before they move downstream.

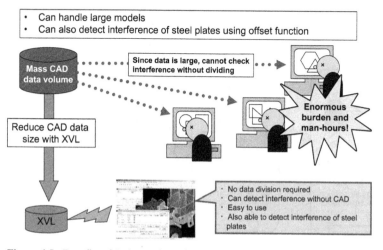

Figure 4.5 Benefits of design review with XVL studio (2)

4.2 3D Parts Lists

The preparation of parts lists is a common and standard practice for managing parts information in the manufacturing industry. Parts lists provide parts information and shape information such as structural information (*i. e.*, component parts making up the part), delivery time for each part, product specifications, supplier, quantity, *etc*. Normally, parts lists are complied by summarizing part attribute information using a word processor or spreadsheet. As shown in Figure 4.6, they are generally output on paper, annotated with notes and hand drawings, and distributed to related parties.

The use of lightweight 3D data for all of this changes the whole situation. Figure 4.7 shows a parts list with 3D shape data. Lightweight 3D shapes can be linked to parts information and shared over the network. Figure 4.8 shows a method of automatically generating such 3D parts lists. XVL tools can automatically combine parts information in Comma Separated Value (CSV) form with 3D shape information in XVL to produce a 3D parts list which can be shared on the network.

Figure 4.9 summarizes the benefits of 3D parts lists. Automatic generation of the parts list eliminates the need to draw illustrations and paste them on the parts list by hand. It also eliminates the need to match parts and part data, and all matching errors that may occur in the process. Also, attribute data can be added to 3D parts lists and the parts lists can also be output on paper. Furthermore, 3D parts lists can be easily updated to track design changes.

Some companies use the same technique to create 3D parts libraries based on XVL. They search for shapes by manufacturer name or part name, and check these shapes using 3D data. As many standard parts have similar shapes, often it is

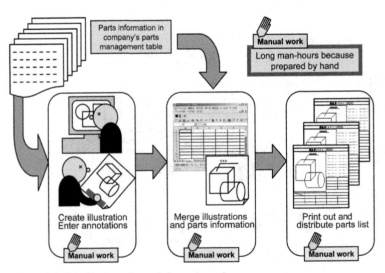

Figure 4.6 Distribution of parts information using paper

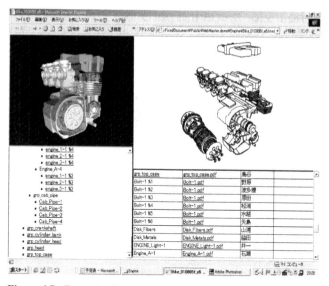

Figure 4.7   Example of 3D parts list using XVL

difficult to identify the correct part from 2D images only. 3D XVL makes part identification much more accurate. In the past it was expensive to build and maintain such parts libraries; however, with the method shown in Figure 4.8 3D parts libraries can be created more or less automatically.

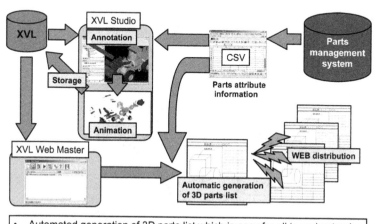

* Automated generation of 3D parts list which is easy for all to understand
* Company-wide information sharing using lightweight 3D data

Figure 4.8   3D Parts list after XVL implementation

- ## Reduction of man-hours for preparing parts list

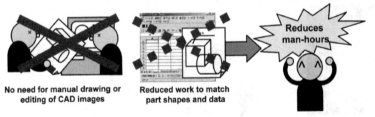

No need for manual drawing or   Reduced work to match
editing of CAD images   part shapes and data

- ## By building the appropriate system...
  - Provide desired Look and Feel of parts list according to requirements of department
  - Automatic assignment of annotations in parts list to XVL
  - Output in table format of parts list for use in offline environment on paper
  - Automated update of design changes

Figure 4.9   Benefits of 3D parts lists

4.3 3D Parts Catalogs

Many companies use parts catalogs to provide parts information for maintenance and procurement. Parts catalogs typically specify parts structure, shape, and attributes. As shown in Figure 4.10, parts catalogs were usually prepared on paper

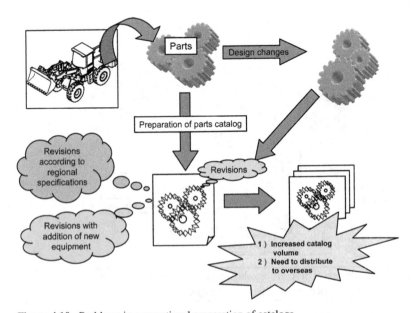

Figure 4.10   Problems in conventional preparation of catalogs

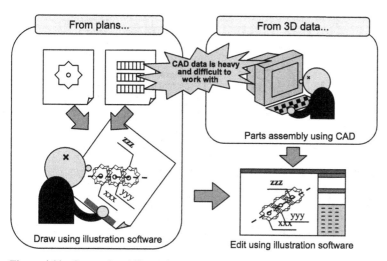

Figure 4.11   Conventional illustration method

and distributed to sales and maintenance people. Global companies also needed to translate or reformat them for distribution to their overseas partners. And they had to be redone every time design changes were made, further increasing costs. Figure 4.11 shows the traditional methods of preparing illustrations for parts catalogs. In one method the illustrator envisions the 3D shape from surface drawings, and draws it by hand using illustration software. Since parts catalogs can contain dozens, hundreds, or even thousands of parts, creating the illustrations by hand requires tremendous labor and time.

In another method, the 3D shape is disassembled using CAD, and image data is created from there. This avoids the illustrator having to visualize the 3D shape from only surface data. But again, this is not practical in the collaborative manufacturing industry where there are hundreds or thousands of parts. Then there is the method where the actual object to be manufactured is obtained, disassembled, photographed with a digital camera, and then an illustration is made from the photograph. However, with large objects like cars, the disassembly time itself will take several weeks, and for even larger objects like cranes, this method is not feasible. It might not even be possible to obtain the actual object. So these traditional methods all have their respective problems.

Figure 4.12 shows an illustration method using lightweight 3D data. Disassembling lightweight data is much easier than disassembling CAD data into parts. By using the method shown in Figure 4.8, it is possible to automatically create parts catalogs directly from the lightweight 3D data. These parts catalogs can then be distributed on the network, on CD, and even printed.

Figure 4.13 summarizes the benefits of this mechanism, which include lower costs for those preparing parts catalogs and better comprehension for the readers of the catalogs. In addition, the data can be delivered worldwide simultaneously over the web. One company who actually introduced this mechanism says that 3D

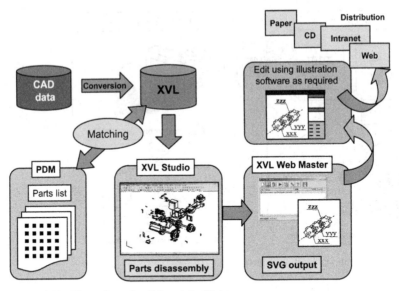

Figure 4.12 Illustration using lightweight 3D data

manuals are clearer and easier to understand intuitively and they are becoming the main source of reference by many of their employees. Employees were provided both paper and network-distributed manuals, and most preferred the digital manuals. Given that not all users are connected to networks, this company distributed paper manuals as standard. However, they found that once a user becomes familiar with the convenience of 3D, they start to prefer digital 3D manuals. This illustration method can be applied not only to parts catalogs, but to the service manuals

- Sharing of Parts Catalog Using 3D Data
  - Even if the parts catalog is eventually printed, the web is generally used for distribution

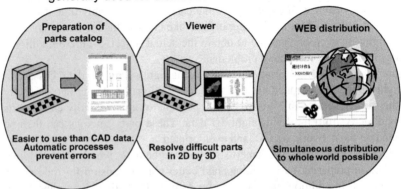

Figure 4.13 Benefits of using lightweight 3D data

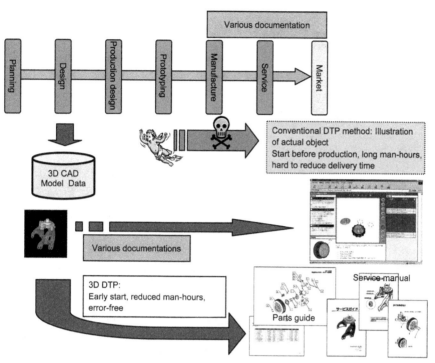

Figure 4.14 Comparison of documentation processes

and in-house technical specifications as well. Traditionally, the preparation of illustrations was started only after the product was completed. But these methods were expensive, had tight deadlines, and produced inconsistent quality. These problems can be resolve by generating lightweight 3D data immediately after completion of design, and using it to automatically generate illustrations.

Creating illustrations from 3D data eliminates the need to produce them manually. It allows the illustrations to be started immediately after design completion. As the illustrations themselves are automatically generated on the PC, the quality is consistent even between different designers, and costs can be cut. As shown in Figure 4.14, changing the documentation process can provide dramatic improvements to quality, delivery time, and cost.

4.4 Animated 3D Visual Manuals

There is a strong trend toward including 3D animations in manuals instead of 2D illustrations. 3D animations are often much easier to understand. It is always a challenge to pass manufacturing knowledge on to new workers, and with the aging of the workforce and increasing globalization of manufacturing, this challenge is only

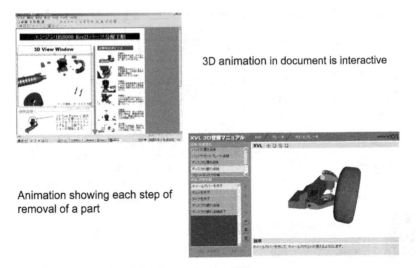

3D animation in document is interactive

Animation showing each step of removal of a part

Figure 4.15 Examples of visual manuals using XVL

going to increase. Animated 3D visual manuals can solve this problem by providing universal instructions that are easy to follow. Figure 4.15 shows two examples of visual manuals with 3D animations. These 3D manuals show animated work procedures, and can be used as assembly instructions and maintenance manuals.

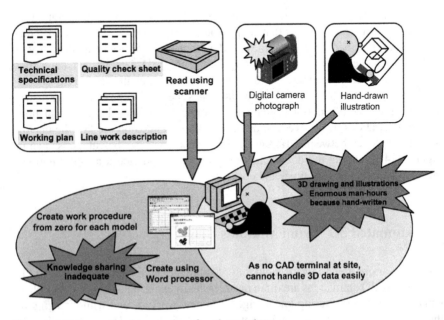

Figure 4.16 Conventional preparation of work procedures

• Work Supervision of Inexperienced Staff by Experienced Instructors

Figure 4.17 Problems of conventional methods

How are conventional non-3D manuals made? Figure 4.16 shows an example. The authors gather the materials they need, scanning technical specifications and work drawings, taking digital photographs, and drawing illustrations by hand. Then they assemble the manual and write the instructions using a word processor. They usually do not have access to 3D CAD systems or data. Figure 4.17 shows the drawbacks of this method. The manual is used by experienced instructors to teach new workers. With more and more Japanese firms advancing into foreign markets, the lack of guidance instructors for overseas workers is a major problem. Moreover, manuals consisting only of words and pictures are not clear to newcomers because such paper manuals are not intuitive.

Figure 4.18 shows the benefits of visual manuals using XVL. Authors are able to directly use 3D data, eliminating the need to draw illustrations or take photographs. 3D manuals allow users to zoom in on areas of interest, view hidden parts from other sides, and thoroughly review important animations. 3D manuals give a greater level of understanding to users. 3D manuals are intuitive and enable employees to teach themselves. And as long as the data is lightweight 3D, the manuals can be distributed online. With delivery deadlines growing increasingly short, it is increasingly important to train for assembly even before the product actually exists. 3D manuals help such training by shortening the instruction time and increasing the trainees' comprehension and retention.

So if 3D manuals are so great, what is the drawback? In the past it has taken a lot of time to create the 3D animations. But new versions of animation software are changing this. For example, with XVL all you have to do is tell the computer the assembly order and it will compute all the animations automatically. You can then touch them up if you like, but you do not have to. Software like this is making it fast and easy to create 3D manuals.

• Sharing of Visual Manuals Using 3D Data on Web

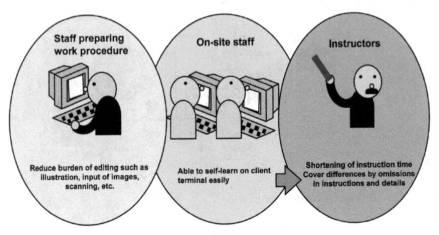

Figure 4.18   Benefits of visual manuals using XVL

4.5 Sharing CAE Analysis Results

Most companies use Computer-Aided Engineering (CAE) tools during the design process. CAE tools can simulate forces, such as stress or temperature, on the model without having to build the actual product. Although some recent CAE systems have become easy enough for the designers themselves to use, most companies hire specialized analysts to analyze the 3D models designed by the designers. CAE

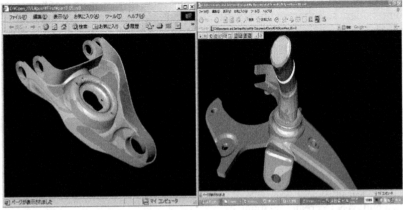

148 Kbytes with XVL   138 Kbytes with XVL

Figure 4.19   Example of CAE data converted to XVL

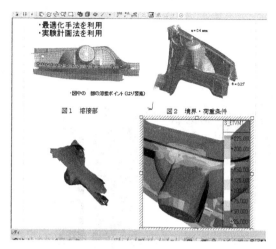

Figure 4.20   Example of CAE data converted to XVL

systems are generally more complex and costly than CAD systems. Furthermore, CAE analysis generates gigabytes of data that can only be displayed using expensive analysis software. Analysts use these results to prepare analysis reports, usually by pasting 2D images captured from the screen. The rest of the data is often discarded. If the designer wants to enlarge some detail or view the results from a different point of view, they need to analyze and calculate again.

One means of resolving this is to convert the analysis data to XVL. The analysis results can be expressed as color contour maps on the geometry. For example, by drawing areas of large stress in red, we can tell at a glance where stress concentrates. Figure 4.19 shows an XVL contour map of analysis results. XVL contour maps can express gigabytes of analysis data in a few hundred kilobytes. Furthermore, with XVL it is also possible to add 3D annotations and hyperlinks, making it an effective means of sharing accumulated design knowledge. Figure 4.20 shows lightweight 3D models in a format that can be distributed on the web. Sharing these analysis results, which were formerly available to selected individuals, throughout the company turns this information into corporate-level knowledge. Young designers are able to learn when problems occur in design, by sharing such analysis results with other data.

4.6 Sharing CAT Measurement Data

Do manufactured products always exactly match the 3D CAD models that specify them? In theory they should, but in reality manufacturing processes are not infinitely precise. In the case of the sheet metal, even if it is machined precisely, it eventually bends due to gravity. The designer must, therefore, take into consideration parts precision and environmental conditions when specifying the model. So how can the

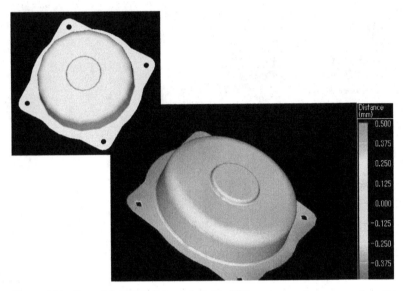

Figure 4.21 Example of CAT data converted to XVL

final product be tested to see if it matches the specified dimensions? The anwer is to use a Computer-Aided Testing (CAT) system. This system compares the final product to the original 3D CAD models to check the manufacturing accuracy.

A CAT system measures the physical product using a point capture device. There are two types of point capture devices: contact devices that actually touch the object and non-contact devices that measure the object using lasers. Lately, the non-contact type has become more popular because it can measure large volumes

- Sharing of comparison data between product and CAD data
- Global sharing of manufacturing knowledge

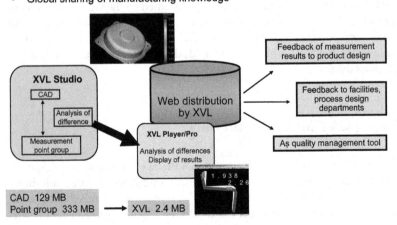

Figure 4.22 Benefits of sharing CAT analysis results

of data in a short time. The measurement results make up a massive 3D point cloud. Comparing this point cloud to the original 3D CAD data used to require expensive analysis systems. But, just like the CAE case described earlier, this analysis can now be performed using XVL and can produce a color contour map. The results can then be shared throughout the organization. Figure 4.21 shows an example of CAT data expressed in XVL. The colors correspond to differences between the 3D model and the actual product. This diagram shows clearly where the final product deviates from the original specification.

With this method it is possible to raise the manufacturing accuracy of all plants – even those that are overseas. For example, the manufacturing process for a part can be worked out at a single plant and then exported to other facilities. They should then be able to produce the same level of quality. As shown in Figure 4.22, an XVL-based CAT system ensures high quality in manufacturing by reporting problems back to design to enable them to further enhance design quality. This is only possible with lightweight 3D data.

4.7 Collaborative Design Using Lightweight 3D Data

Another feature unique to lightweight data is the use of 3D data for simple communication. As shown in Figure 4.23, 3D data can be used effectively by all departments related to design and also between partner companies. One example is the

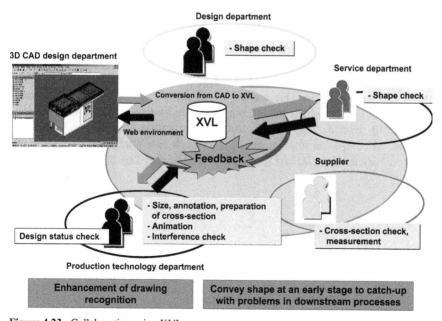

Figure 4.23 Collaboration using XVL

exchange of data between designers and engineers. In the past, a designer in a different location had to send 2D drawings to the engineer, who would then build a 3D CAD model and bring it back to the designer for discussions.

With XVL, the designer is able to send a lightweight 3D model to the engineer, who visually checks it, thus reducing the number of meetings needed. Also, the designer can send XVL models as soon as the initial design is complete, further enhancing communication with the engineer. One of the greatest benefits of XVL is that it runs on a standard PC. For example, when demonstrating the design concept, the designer does not need to go to the CAD room where the engineer is stationed. All he or she needs to do is go to a meeting room with a notebook PC. Then he or she can demonstrate 3D animations of the product to any audience – design, engineering, manufacturing, sales, marketing, and maintenance.

3D animation allows those who are unable to picture the design from drawings to understand the actual product. Data used in this way can also be used for presentations outside the company and for describing marketing purposes to clients. As the data is very simple and easy to understand, clients know instantaneously if they will like the product. Marketing quality is therefore high, compared to verbal product image descriptions.

Lightweight 3D data is also useful for communicating with suppliers such as mold manufacturers. For example, if the outsourcer and mold manufacturer do not use the same CAD systems then it may not be possible to display each other's data. Mold manufacturers with multiple clients simply cannot afford to have different CAD systems for each of them. XVL provides a solution by allowing simple shapes to be shared between the mold manufacturer and outsourcer. Each party can convert their data to XVL and email it to the other. Each party can then review and annotate the data and send it back. This process can be used, for example, to check the parting lines on the model.

This chapter has introduced several applications of lightweight 3D data, from simple methods of checking 3D shapes to enhance communication, to further increasing information content by combining 3D data and manufacturing attribute data or adding animation. One leading-edge example is the method of sharing CAE and CAT data, which many companies are already using. The following chapters will describe case studies from three countries, Japan, the USA, and Germany, of companies implementing XVL. Each case study will describe how and where they use 3D data and the major issues they have encountered in its use.

Chapter 5
Design Review in Body Design: Case Study of TOYOTA MOTOR CORPORATION

It is said that 80% of the production cost of a product is determined in the design stage. Considering the total costs of design and manufacturing in the product life cycle, it is therefore extremely important to design high quality from the beginning. At TOYOTA MOTOR CORPORATION (TOYOTA), body design quality is enhanced by performing design review (DR) using XVL. This chapter discusses the actual uses of 3D data in DR as described by Mr. Junichi Harada, general manager of the body and electronic parts design division at the Lexus Center. Mr. Harada says he is concerned about depending too much on 3D, so he is careful about how he uses it. DR is one area where he has found clear benefits from using 3D. (Note: The following is based on the keynote speech by Mr. Harada at XVL Solution World held on May 11, 2006.)

5.1 Why is Design Review Necessary?

In order to respond to consumer demand more quickly, automotive manufacturers are updating their models more frequently. This shorter product change cycle time is causing major changes to the body development process. More and more companies are switching from progressive development, where production preparations are started only after design has been completed, to concurrent development, where design and production planning proceed simultaneously (Figure 5.1). TOYOTA calls this process simultaneous engineering.

Simultaneous engineering is the method whereby development creates an initial design and then development and production work together to refine the design and to prepare for production. In the beginning, a design is developed that meets the design goals. If problems are detected during the design process, design changes are made by the designer in charge in real-time. Because the designer designs using 3D CAD, the design changes are incorporated into CAD data on a daily basis. Ideally, at the same time drawings that meet the requirements are finalized and die and mold making is started.

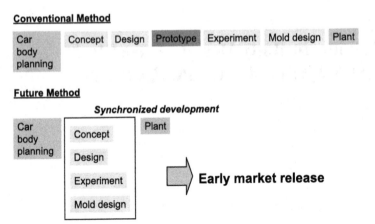

Figure 5.1   Conventional design methods are gradually changing to simultaneous engineering methods

In the simultaneous engineering environment, both the parts that a designer is in charge of and neighboring parts change all the time. In some cases, the neighboring parts change before one even realizes it. This is because the respective designers change design according to the local design criteria. As a result, interference can occur between parts from different design departments. The automobile industry, which uses collaborative manufacturing, produces products that contain several thousand parts. So interferences can occur very easily even with the simplest of shape changes. Such interferences can trigger major problems in the downstream manufacturing process. The later a problem is detected, the greater will the costs be for changing the design, changing manufacturing facilities, and changing dies and molds. Ultimately such mistakes can even delay the delivery time. Thus even small design mistakes can have a dramatic influence on downstream processes. One means of resolving this problem is DR. DR can detect interference problems early in the process. But DR is not a one-shot process; it is important to repeat DR since design changes to resolve problems can themselves cause new interferences. The more simultaneous engineering is implemented to shorten product development time, the more new problems will be created. The key to success is resolving these problems early in design, and one way to do that is to develop a DR system using 3D data.

5.2 Design Review Using XVL

TOYOTA uses Lattice's XVL Studio Pro in their DRs. In the past, they carried out DR according to the following procedure:

1. Since CAD systems cannot display very large assemblies, remove all unnecessary parts
2. Generate cross-sections of the reduced CAD assemblies
3. Visually check the cross-sections to find interferences

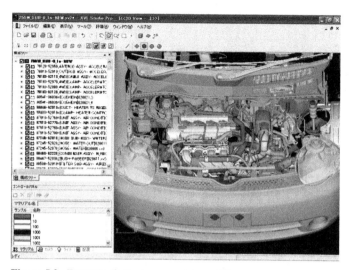

Figure 5.2   Example displaying mass data on PC

Interferences can be detected by checking the cross-sections. However, this method requires the designer to visually check large numbers of cross-sections, resulting in a tremendous burden on the designer. Also, there are the risks of failing to generate the right cross-sections and of failing to spot interferences. Furthermore, this burden is heaviest for the designer during the busiest times of the year.

XVL Studio Pro is a design review tool that resolves these problems. It is able to:

1. Display and process a complete automobile at one time on a PC (Figure 5.2)
2. Automatically calculate interferences between all parts, and display cross-sections of interferences in real-time (Figure 5.3)
3. Automatically generate reports of interference results including pictures and diagrams of the interferences (Figure 5.4)

At TOYOTA, DR reports are used as overall management sheets, and all problems are investigated until they are completely resolved. The overall management sheet has a check column to indicate at a glance whether a problem has been resolved or not. In actual operations, two types of DR are carried out: DR in design stage and DR after drawings are re-drawn after the final design. DRs in design stage are carried out in order to prevent problems from getting out of the design area. New problems emerge every day during the simultaneous engineering process, and repeated DRs are an effective means of resolving them before they magnify. As shown in Figure 5.5, problems are discovered and resolved with each successive DR.

The final DR, which occurs after the final drawings are created, is the final design check. It is performed by a third party. At that point all problems must be resolved. XVL Studio Pro minimizes the time required for the final DR, reducing the burden on the designers and enabling them to focus on essential and innovative design work.

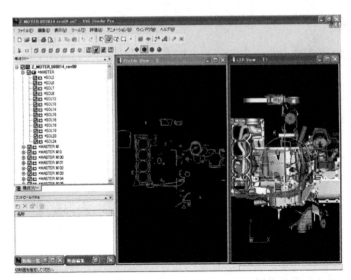

Figure 5.3   Real-time display of cross-section of interference

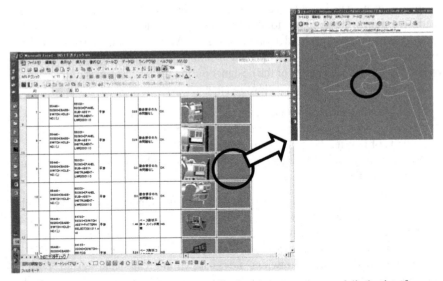

Figure 5.4   Interference results report preparation function, can enlarge and display interference

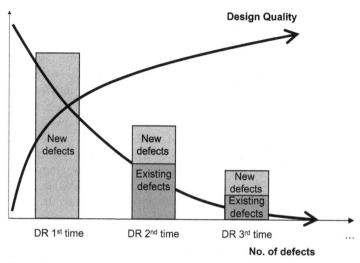

Figure 5.5   Repetition of DR decreases number of defects

5.3 The Actual Design Review Process

At busy design sites it is usually difficult to introduce new business processes and tools. In order for new processes or tools to be accepted, they must provide clear benefits and they must not increase the burden on the designers. One effective way to introduce lightweight DR to a busy design team is to set up a DR Preparations Team to collect the required 3D data, convert it all to XVL, and then demonstrate the use of XVL Studio Pro at the DR site. This will enable the designers themselves to see how XVL Studio Pro helps prevent interference problems that could not be discovered any other way. Once they understand the benefits, they will embrace lightweight DR.

At TOYOTA, DR is currently carried out in two steps; sheet metal DR and DR of all parts. Sheet metal DR checks the interference between sheet metal parts. DR of all parts checks the interference between all parts. The DR of all parts is intensive. Meeting schedules are determined for each body part (door, front body, *etc.*) to check interferences in detail. All related staff are invited to each meeting. Before each meeting the DR Preparations Team performs an automatic interference check using XVL Studio Pro to find problems. During the DR meeting the staff resolve all of the problems on the spot. As simultaneous engineering progresses, interference problems occur every day. XVL Studio Pro can automatically perform comprehensive interference checks on very large assemblies and generate reports. The problems are managed on overall management sheets and resolved one by one in follow-up meetings. All parties involved are gathered and all problems are resolved on the spot. This process saves a lot of time for the designers and manufacturers, enabling them to spend more time on creative work such as developing and manufacturing new products.

5.4 Applications and Development of Design Review

As designers begin to experience the benefits of XVL Studio Pro, they start to become more and more positive and enthusiastic about using it for a diverse range of purposes. For example, they want to use it not only for resolving interference problems, but also for solving many other kinds of problems. There are many applications for XVL beyond interference checking. XVL design reviews can be used for determining the optimum arrangement of cables and wires, reviewing assembly efficiency, checking the clearance between movable parts, preventing abnormal sounds, and reviewing product appearance. XVL allows everyone present to check these problems together visually, determine what is wrong, and find a solution on the spot. For example, consider the problem of determining the optimum arrangement of a wire harness in an instrument panel. Wire harnesses are arranged in a complicated manner in an instrument panel, and XVL Studio Pro can be used to check the arrangement. CAD is unable to display such large-scale data, but XVL is able to display all of it at once, thus enabling design verifications. Sometimes, however, as it is difficult to visualize the bent and twisted portions of the wire harness on the computer even with 3D. In these cases DR is carried out while comparing the actual parts. XVL can also be used for problems related to the assembly procedure. Sometimes the shape or position of a part can cause problems for assembly or repair. These assembly or repair problems can often be resolved by slightly moving or reshaping the part. By displaying all the related parts by XVL, everyone present at the DR can identify the assembled state of parts, thereby accelerating the problem-solving process.

This gathering of related parties and visual inspection of large assemblies displayed in XVL allows solutions to be reached easily. In the past, even simple problems took time to resolve if they concerned different divisions. Other applications of XVL DR include building construction and appearance review. During construction basic planning, XVL is used to determine if the design follows the building concept envisioned by the designer. 3D visual display enables the designer to quickly see if parts assembly and positions differ from the original plan. In appearance review, XVL allows models to be viewed from all directions instead of just the limited 2D drawing views. For instance, with 2D drawings it is difficult to review the appearance of the lamp at the corner of a car through the window glass, but 3D display allows the appearance to be checked from all angles. If the designers are able to identify the problem, they will be able to come up with resolutions naturally. These examples show that there are many different ways of using XVL in DRs. Other design verification applications are being developed as designers brainstorm together and identify new targets. The important thing in these cases is to clear what is to be checked during the DR. Interference checking is just the beginning. The applications of 3D DR using XVL are as limitless as the creativity of the designers that use it.

5.5 Advantages and Disadvantages of 3D Design

According to Mr. Harada of TOYOTA, the basis of design, that is the objectives (intentions) of design, basic dimensions, rationale of the design, *etc.*, are decided by the designers themselves. XVL is merely a means to enhance the degree of design completion. It is still up to the designer to enhance basic design. The designers must draw up basic design plans. So designers still need to understand the basics of design and manufacturing.

Recently, there is a trend for young engineers to rely on 3D data as it is simpler. For example, there are more and more cases where the shape is ready, but the overall structure and design intent are not clear. According to Mr. Harada, this may be due to the fact that designers tend to rely on XVL and think that they can review later with XVL. Mr. Harada says that the core of design can essentially be expressed only on drawings; this is the product concept envisioned by designers (Figure 5.6). According to Mr. Harada, 3D CAD shape data is merely the result of changing this "core of design" to shape using CAD. Of course, 3D shapes are convenient. It is easy to check shapes and easy to communicate between development members using 3D. However, designers must express the performance and functions of design on drawings and guarantee this. Supervisors of design divisions have the important responsibility to pass down the "core of design" and spirit of manufacturing to young designers. This is an important principle for Japanese manufacturers.

In the actual design process, it is necessary to make full use of the advantages of both drawings and 3D data. Even with DRs, it is better to carry out both DRs based on drawings and DRs using XVL. In drawing-based DRs, the "core of design" can be checked and the overall concept identified, and sense of scale confirmed. The advantages of DRs using XVL include spatial effects, verification of minute parts, and the ability to check problem parts on the spot. By differentiating use of either according to requirements and integrating the advantages of

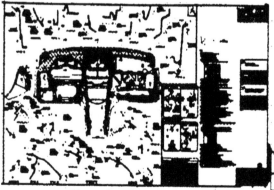

*Source: Japan Automobile Manufacturer's Association

Figure 5.6   Heart of design, basics of design specified by drawings

both methods, the degree of completion of drawings can be enhanced in the true sense of the word, which means degree of completion of design also improves.

Of course, the "core of design" prescribed on drawings is converted to 3D shapes by CAD, and by converting 3D shapes to XVL it can be used in downstream processes and by suppliers. 3D data is used as a means to convey information. In the future, the trend will be to use 3D data instead of drawings, but as long as 2D drawings are the final authority, the will continue to convey the "core of design." Mr. Harada says that designers must learn to thoroughly identify what is the core of design before actually proceeding with the design of 3D shapes. And in the ideal design process, the design concept envisioned is expressed on CAD as 3D shapes.

5.6 Two Goals of Using XVL

There are two ultimate goals in using XVL. One is to reduce the time of trivial work such as interference checking, which will free designers to concentrate on creative work such as design. The second goal is to maximize design quality by combining the merits of both drawing-based DR and XVL-based DR. This uncovers problems early in the design process and prevents them from reaching downstream, thus reducing costs and delivery time. However, no matter how useful a tool is, it means nothing if it is not used. Tools must be used efficiently to achieve the goals in design. Simultaneous engineering has the drawbacks of creating more interference problems in design and increasing the amount of rework during design and manufacturing. The greatest goal of DR is therefore to completely eliminate such unnecessary work. This will result in improved quality and faster delivery time in simultaneous development. The importance of DRs is ever-increasing in the integral manufacturing-type industry, which creates functions and performance by the combination of numerous parts.

Chapter 6
NIKON: Use of 3D Data as a Communication Pipeline

NIKON Precision Inc. is a manufacturer of semiconductor fabrication exposure devices (steppers) for manufacturing semiconductors and liquid crystal display (LCD) exposure devices for mass producing LCDs. These devices consist of numerous parts and are typical collaborative manufacturing-type products. The use of 3D data can therefore significantly improve design efficiency at this company. The use of 3D data can also improve the efficiency of downstream processes. And although the effective use of 3D data in downstream processes still requires considerable trial and error, XVL has become an indispensable means for communication at the NIKON group. This chapter contains an interview with Mr. Kiyotaka Yamamoto, the head of the precision machinery department at NIKON Precision Inc., about their endeavors to incorporate XVL in their design and manufacturing activities.

6.1 Environment of Semiconductor Fabrication Devices

The semiconductor fabrication device is a machine for mass producing semiconductors required for digital equipment. Semiconductors are indispensable to automobile control devices, PCs, cellular phones, *etc.* As shown in Figure 6.1, this device has an extremely complicated mechanism, and consists of up to 100,000 parts in total. Product development time is usually long, over one year. In order to manufacture semiconductors which are growing more and more complex, the manufacturing systems themselves need to be high precision, and new manufacturing technologies are required to keep pace with the growing precision. Also, competition with rivals in the same industry is driving shorter and shorter product cycles.

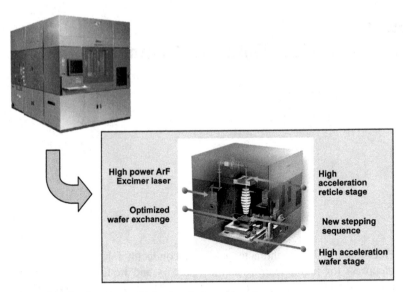

Figure 6.1   Semiconductor fabrication device and mechanism

In this severe environment, NIKON's development department encountered a huge obstacle. In order to design products which continue to grow massive in scale, they were required to divide the design responsibilities. The design engineers for different areas became highly specialized, and they were committed to maximally optimizing their individual parts. As a result, there were very few design engineers who were able to oversee the entire development process, so design reviews (DRs) to grasp the whole picture had little substance. It was therefore difficult for NIKON to optimize design which took into account the whole picture and motivate design engineers.

That was one problem. Another problem was that at that time the company had no CAD facilities in its downstream processes. Even though 3D data was available, they had yet to be able to use the 3D data effectively. Work could only be carried out on drawings. The key to resolving these problems was to restructure the business processes and make them more efficient by incorporating 3D design. This consists of two aspects: design process innovation by 3D design and use of lightweight 3D XVL data in post-processes.

6.2 Design and Manufacturing Process Innovation with 3D Data

Figure 6.2 shows the applications of 3D data at NIKON. In their design department, 3D CAD is used by more than 100 staff members to design each functional part. CAD data is directly used when there is a need to prepare tool sheets and assembly check sheets simultaneously with design work. For other applications in production and manufacturing, such as shape verification, manuals, *etc.*, the CAD data is con-

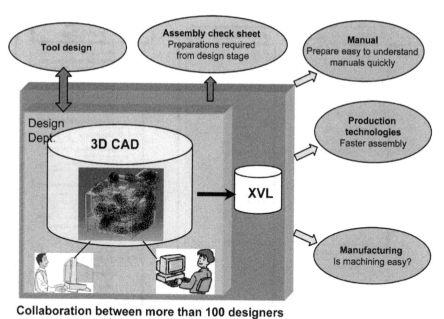

Collaboration between more than 100 designers

Figure 6.2   Use of 3D data in design and post-processes

verted to XVL and used. Figure 6.3 shows the areas using 3D data at NIKON. In their design department, CAD is used to perform design. Even in departments requiring CAM data, such as machining departments, CAD data is used as it is.

On the other hand, lightweight XVL is used in production management, in downstream processes such as logistics and development, and in the preparation of presentation materials. At NIKON, 3D data was first used by the design department. In 2003, 3D CAD was installed with the goal of verifying the whole product in 3D. 3D data was also used during meetings to help motivate design engineers.

As semiconductor fabrication devices are made up of 100,000 parts, design engineers need to cooperate with each other in design work. Since CAD is unable to display the whole semiconductor fabrication device at one time, in the past each designer would review only the parts they controlled. For DRs, in the past, design engineers would prepare 2D drawings and explanatory full scale drawings. Now DRs are performed directly using 3D data. The design department no longer needs to prepare explanatory drawings in order to explain their designs clearly. The use of 3D designs therefore encourages and motivates participating staff to a great extent in every way.

Figure 6.4 shows the adoption curve of 3D CAD in manufacturing. Given that 3D CAD is not easy to use, in the beginning NIKON had specific CAD operators instead of design engineers using the 3D CAD systems. However, design engineers gradually started to use 3D CAD systems themselves, because through DR they experienced and enjoyed the advantages of 3D CAD. Eighteen months later the design engineers at NIKON were all using 3D CAD.

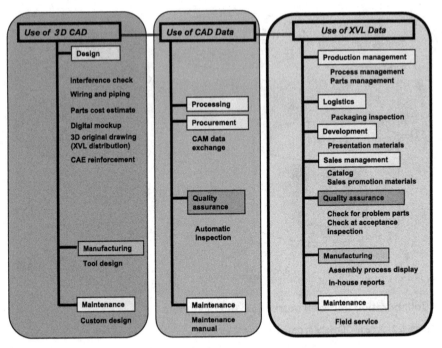

Figure 6.3   Promotion of use of 3D data by department

Initially PCs and projectors were distributed to the design department during meetings and DRs in order to view the 3D data. Within six months 3D data was used in all design meetings, and within two years it was used throughout the company.

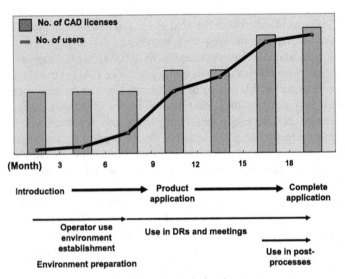

Figure 6.4   Adoption of 3D CAD in the design department

6.3 Difficulties Using 3D Data in Downstream Processes

NIKON encountered difficulties using 3D data in downstream processes. 3D data was shared between the manufacturing and design departments. However, the company found that the downstream staff would not use 3D CAD if it was not convenient or necessary, and especially if its usefulness was not clear.

The downstream departments are divided according to function, such as the assembly department, the publications department, *etc.* This made it difficult for those promoting the use of 3D data as they could not find advantages common to all departments and did not know how to promote the use of 3D data to each department. The breakthrough was to familiarize the downstream departments with 3D data through DRs. By carrying out DR based on 3D data, the manufacturing department would ask the design department for specific 3D data. Manufacturing staff found that they often wanted to use 3D CAD but did not know how to. They needed a lightweight 3D solution. Next, NIKON had to choose the type of lightweight 3D data to use. Given the advantages of small data size, free viewer, and easy conversion of 3D CAD data, they chose XVL. But upon actual use, they immediately encountered problems. The semiconductor fabrication device had too many parts. It took five days to manually perform XVL conversion of 70 units with 70,000 parts. This meant it would be impossible for them to achieve their initial goal of supplying design data to the manufacturing stage in a timely manner.

The solution was to develop an automatic conversion system. In just one month they added XVL conversion into their CAD data management system. The new system automatically converts data to XVL for each part and assembly, and can convert 70,000 parts overnight. With this system, they are able to send data to downstream processes as development reaches each level of completion. Finally, their design department was able to share 3D data with the manufacturing process in a timely manner.

6.4 XVL's Role as a Communication Pipeline

At NIKON, XVL is gradually establishing itself as the cornerstone of communication unimaginable in the past. In the beginning, 3D data was used for meetings and documents. Figure 6.5 shows a meeting discussing product improvement using XVL data. XVL is easy and convenient in meetings of downstream processes. Figure 6.6 shows an example of reports and procedures. Previously drawings were sampled, converted to a 2D drawing interchange file (DXF), and pasted in to documents. After the actual part was manufactured, it was photographed using a digital camera and the picture was used to create drawings. Now, 3D XVL models are simply pasted in to Microsoft Word and Excel. This has sharply reduced the time and work required to prepare documents, and also enabled manuals to be created immediately after the design is completed and the 3D data is available.

Figure 6.5   Meetings using XVL for product improvement

By using 3D data, NIKON has been able to cut costs by 30% and shorten delivery time by 15% in preparing manuals and reports. NIKON also uses XVL to find similar shapes in design data, reducing the time for shape recognition by 10%. At NIKON, XVL has become an indispensable tool for communication. Figure 6.7 shows that information can now be accurately and efficiently conveyed by 3D data not only inside the company, but with other NIKON group companies and with partner companies.

At present, NIKON says they are advancing yet another step forward. As the adoption of 3D design has sharply increased the workload of design engineers, they are currently attempting to reduce this cost by replacing some of the 2D drawings with 3D data (Figure 6.8).

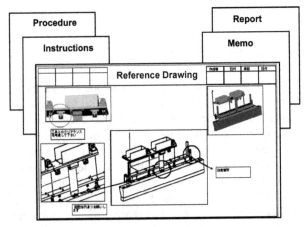

Figure 6.6   Example of use of XVL in specifications and procedures to enhance communication efficiency

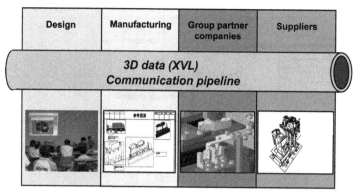

Use of 3D outside the organization, efficiently communicating design intent

Figure 6.7 XVL is a MUST tool for efficient communication

In the past, they prepared assembly drawings by combining parts component information (component lists) and several drawings. With large-scale products, in order to convey information in a clear manner, design engineers had to make many drawings, which usually took one or two weeks even after the designer had completed the design using 3D data. By improving the display and operational methods of the XVL viewer and making it easier to check the attached position of parts, the company was able to replace these drawings and use XVL to convey the positional information of parts.

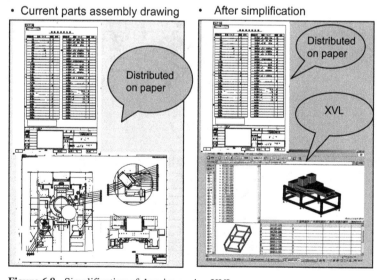

Figure 6.8 Simplification of drawings using XVL

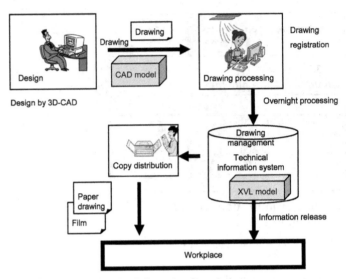

Figure 6.9 Flow of design information from design to post-processes

Figure 6.9 shows the flow of information from design at NIKON. Design is performed by CAD, and at the point design is completed, component lists and CAD models are released. The part data and models must be combined in a process called drawing registration. With XVL the data is automatically processed at night, and the part data and XVL model are linked and released to related departments.

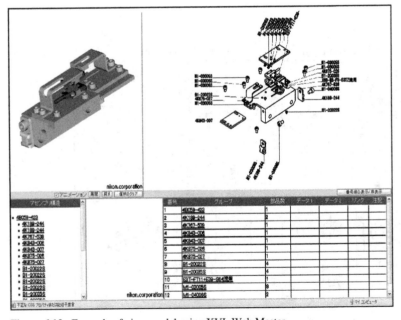

Figure 6.10 Example of view model using XVL Web Master

Figure 6.10 shows a released view model. It was automatically generated by XVL Web Master using one of the basic templates. One outstanding feature of this template is that the parts list and shape are linked, and this has dramatically reduced the number of required assembly drawings.

6.5 Security: A Pending Task

To further promote use of 3D data, NIKON say they need to overcome two tasks related to XVL, namely handling of enormous volumes of data and handling of security. As the use of a semiconductor fabrication device involves massive data volume, high-speed display of large-scale data and high-speed illustration are necessary.

The problem of data display is being resolved by the newly developed V-XVL technology (xv2) by Lattice. XV2 was specifically designed to display large volumes of data quickly and with a small memory footprint.

The problem of security arises from the lightweight nature of XVL. XVL is small enough to be easily sent by email and read by outsiders. So the company has developed an XVL security system linked with in-house authentication systems. Generally, such systems should be installed to ensure security before further promoting the use of 3D data.

In summary, the introduction of XVL at NIKON provided three main advantages:

1. Realization of front-loading by improving efficiency of design process and motivating design staff, and promotion of communication between manufacturing staff and design engineers
2. Sharp reduction of time and costs for documentation by using 3D data
3. Easier and clearer communication between the design and manufacturing departments, partner companies, and suppliers

In particular, NIKON found unexpected benefits from using XVL as a communication pipeline. When the company resolves issues such as security, they will be able to benefit even more from the use of 3D data. NIKON Precision Inc. is steadily succeeding in the effective use of 3D data in their downstream processes.

Chapter 7
YAMAGATA CASIO: Digital Engineering Practiced at Injection Mold Plant and Transfer of Technological Information

YAMAGATA CASIO is the manufacturing subsidiary of CASIO, a company which manufactures watches, digital cameras, cellular phones, *etc.* The business activities of the subsidiary include product assembly for CASIO and an independent parts business which includes the manufacture of precise mold and molded parts. The company introduced 3D CAD/CAM into their parts business from a very early stage, to ensure thorough data use via networks and start the complete digitalization of their manufacturing process. By applying XVL to production instructions and assembly instructions, they have been able to improve production efficiency by 30%. They have also integrated XVL into their process management systems. The following sections are from an interview with Mr. Takaya, Deputy General Manager, Engineering Department, Plastic Device Devision of YAMAGATA CASIO, as of May 2006.

7.1 Digital Equipment Market

The prominent products CASIO makes, like mobile phones and digital cameras, have a short lifespan; four months in the case of mobile phones and six months for digital cameras. Based on the ever-changing consumer need, product assembly manufacturers like CASIO have to shorten development lead time (Figure 7.1).

Set manufacturers such as CASIO are customers of YAMAGATA CASIO. These set manufacturers produce products quickly by carrying out production preparation in one go. YAMAGATA CASIO has successfully shortened production preparation time by adopting 3D CAD data and digital distribution of technological information. In addition, they also have a system to automatically gather data via the network and monitor machine tools and molding machines on the network, thus creating a stable supply of molded parts.

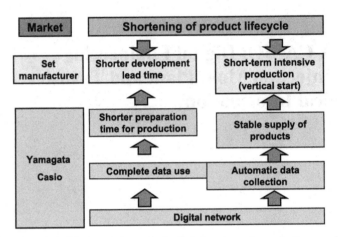

Figure 7.1 Market needs sought by digital equipment manufacturers and response

7.2 3D CAD/CAM and Network

Figure 7.2 shows the transition in 3D design data. The company introduced 2D CAD in 1988, and then CAM/CAE systems. In the first half of the 1990s,

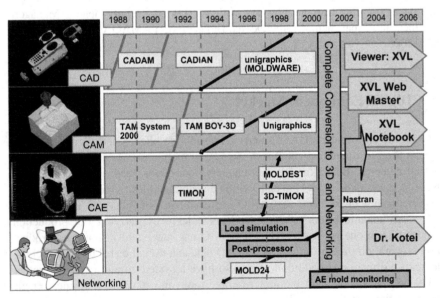

Figure 7.2 Conversion to 3D data and networking in design and manufacturing work at YAMAGATA CASIO

they introduced 3D CAD, linked it to CAM/CAE, and started using 3D data. As a result, in as early as 2000, the whole design department was already using 3D CAD/CAM systems. At the same time, the company also built a system to transfer information via the network. In 2002, the company started to focus on the use of 3D data in design and manufacturing. They began to generate XVL as viewer data and to transfer technological data using such tools as "XVL Web Master" (management tool which links parts list and XVL) and "XVL Notebook" (documentation tool). In addition, they digitally linked CAD/CAM/CAE systems, mold machine tools, and molding machines for monitoring purposes. For machine tools, they networked machining centers and electric discharge machines, and for molding machines, they even linked their auto stockers (machines that pack products into boxes) to the network. If they found any problems in the processing state, the system would mail the situation to the mobile phone of the person in charge.

Figure 7.3 shows the flow of digital data. The company receives product data and specifications from their customers. Based on this, the company models the product, designs the mold, and designs the jigs (tools) using the Unigraphics 3D CAD system.

After this, based on customer requirements, they construct prototypes using 3D. They use the resulting 3D shape data for CAM and CAE. For downstream processes, they convert data to XVL. For example, in the jig design process, they would use XVL to determine the best method of controlling the location of a work piece by a jig for mass production.

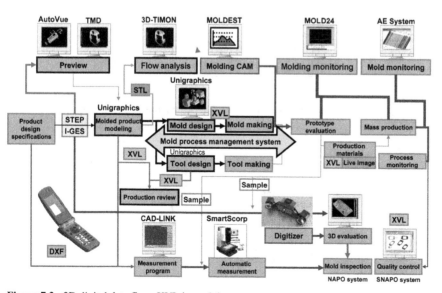

Figure 7.3 3D digital data flow: XVL is used downstream

7.3 Why 3D Design Alone is not Effective

This brings us back to the question, what exactly is design? According to Mr. Ta-kaya, it is embodying and expressing whatever one envisions (Figure 7.4). Using whatever knowledge is available, the designer expresses his or her best vision according to specific procedures. In other words, the designer first decides what he or she wants to make, picks the shape and material, decides how to make it, and expresses this on drawings and specifications. Drawings are therefore for express-ing the intent of the designer, not a means to transfer shape. The design intent that the designer wants to convey is diverse, ranging from shape and size to tolerance, joint state, important dimensions, materials quantity, machining procedure, and finished state, and most of this design intent is conveyed using drawings.

3D CAD data only expresses shape, which means the effects of designing by 3D are limited. This is because without drawings, the designer is unable to convey his or her intent correctly. Drawings therefore still play a very important role in the design process. After completing 3D design, the designer has to make numer-ous drawings. According to Mr. Takaya, at YAMAGATA CASIO they started to use CAD to work on 3D surface models instead of 2D, and then went on to use 3D solid models to implement 3D design. However, even in such an environment, it had not been easy for them to stop using drawings. They have finally been able to realize complete use of 3D data with the evolution of XVL and after the tools for XVL became available, and established a system for transferring the design intent to downstream processes without drawings.

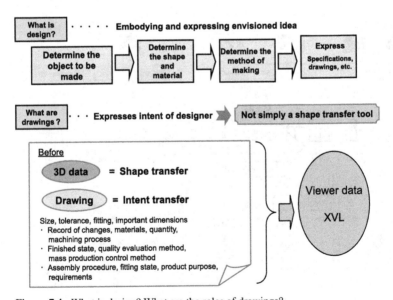

Figure 7.4 What is design? What are the roles of drawings?

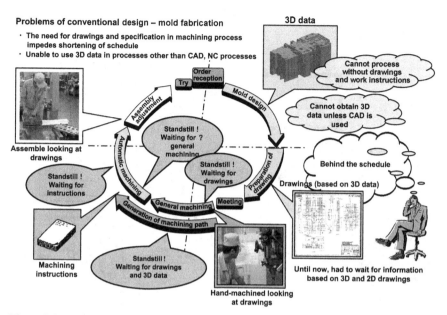

Figure 7.5 Problems of conventional design methods

As Figure 7.5 shows, in previous design and manufacturing processes, YAMAGATA CASIO had been using 3D data and 2D drawings that express design intent because without drawings and work instructions they were not able to carry out their manufacturing work. However, in their design process, they only can start drawing after completing 3D design. The essential problem lay in this drawing process, which hindered them from shortening the manufacturing time.

The introduction of 3D design proved very effective in terms of accuracy and quality. But then again, in order to enhance efficiency of downstream processes, they had to eliminate drawings completely. They found that they were unable to shorten overall lead time just by introducing 3D design.

7.4 Ideals of Design and Mold Fabrication

Figure 7.6 shows the ideals of the process from design to mold fabrication. Ideally, only 3D data should be used from design to all processes at the manufacturing site. As the figure shows, manufacturing problems faced at YAMAGATA CASIO can be resolved by a process management system and without using drawings. All molding information can be centrally managed.

In the design process, previously attribute information was incorporated in drawings. Now it can be input by a CAD system. Custom programs can then extract the attribute information as a table in CSV format. This processing informa-

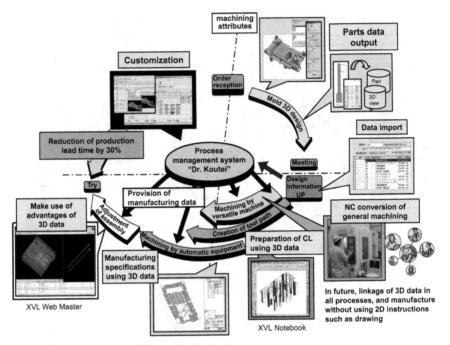

Figure 7.6 Design and mold fabrication in the future

tion and parts information can be integrated with XVL and shared. Furthermore, this sharing can begin at the earliest stages of design. For the downstream processes, status information allowing accurate tracking of schedule changes is provided.

XVL Notebook is used to create processing instructions with 3D data. This tool allows part attribute information from the CAD data to be seen at the manufacturing site. 3D data and attributes linked, and the joint state of mold surfaces, which is important for mold making, can be visually checked. The 3D data also enables the user to understand complicated fittings that would not be clear from drawings.

This system allows accurate information to be transferred even to outside processors. XVL Web Master is used to generate assembly instructions, allowing verification of the part positions and the actual assembly process.

This system helped YAMAGATA CASIO shorten production lead time by 30%, something which was not possible with 3D designs alone.

7.5 Introduction of Process Management System

Another important endeavor was the company's introduction of a process management system (Figure 7.7). This system transfers design information to down-

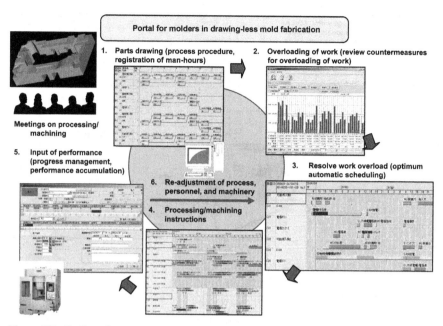

Figure 7.7 Outline of process management system

stream processes, tracks the progress of each process, and manages the workload. They chose the "Dr. Koutei Pro" system developed by CIM Technology, Co., Ltd. "Koutei" means a step in the manufacturing process. Integrating XVL into this system enabled them to fabricate molds without using any drawings. This system contains the processing procedures and required man-hours for each and every part. All the man-hours required for fabricating multiple molds are added up and the points where workload tends to concentrate are checked. Processes requiring excessive man-hours are re-planned to level out the workload.

Actual values are input in this system according to the progress of manufacturing at the company. Totaling up and distributing the results each day provides daily updates to the management plan and enables processes to be rescheduled as needed. As shown in Figure 7.8, operators refer to parts diagrams specifying the rough work procedure per part and the Gantt chart indicating work details to confirm what is to be done that day. Since 3D data of parts is also linked to the process management system, this combination of process management system and XVL allows the manufacturing process, progress, and product data to be checked all at once. For example it shows instantaneously which parts of the mold had problems, *etc.*

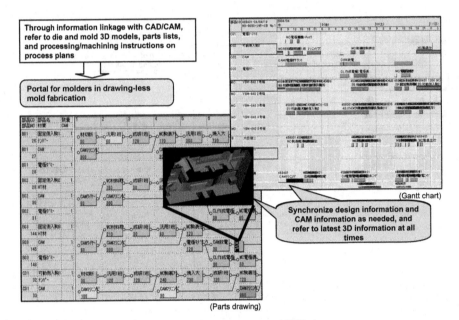

Figure 7.8   Integration of process management system and XVL data

7.6 Changes in Information Transfer Media with Increased Use of 3D Design

Figure 7.9 shows the changes in the design tool at YAMAGATA CASIO, and how, as a result, their media for transferring information have changed. In the 1990s, they changed from drafter to 2D CAD. However, they continued using parts drawings to express most product management information such as dimension, tolerance, and materials. Then in the latter half of the 1990s, they introduced 3D CAD and were finally able to replace shape and dimension data with 3D data. At the same time they changed the method of conveying information. Instead of drawings, information such as design changes and finished state were transferred through technical meetings. However, for mold inspection specifications they had to write instructions for the inspection as 2D drawings and for assembly procedures they had to describe how to assemble the parts on paper. This information was therefore conveyed to downstream processes by the analog means of paper.

Currently, YAMAGATA CASIO is experimenting with 3D digitizer to evaluate 3D products. As shown in Figure 7.10, the 3D digitizer measures the object and produces a point cloud. The point cloud is compared to the original XVL model, and the differences are mapped on 3D data as a color map. The company says this 3D evaluation method allows them to easily find warps and deformations in the final product.

Design Tool	Drafter	2D CAD	2D CAD 3D CAD (Surface)	3D CAD (Solid)	3D CAD (Solid) Viewer
	CAE (Cell)			CAE (Solid)	
Media	Paper drawing	Paper drawing (2D data)	Paper drawing (2D data) SF data	3D data Technical meeting Paper drawing	3D data Attribute Instructions
Shape	Parts drawing	2D data		3D data	3D data
Model Design	Design Mockup		SF data	3D data	3D data
Dimension	Parts drawing			3D data	3D data
Dimensional tolerance	Parts drawing			Technical meeting	Notebook
Record of revision	Parts drawing			Filename	Notebook
Material	Parts drawing			Technical meeting	Notebook
Finished state	Parts drawing			Technical meeting	Notebook
Quantity	Parts list			Parts list	Web Master
Fitted state	Parts drawing			Technical meeting	Notebook
Critical dimension	Parts drawing			Technical meeting	Notebook
Dimension evaluation	Parts drawing				3D Digitizer
Quality assessment	Parts drawing				Notebook
Assembly procedure	Instructions				Web Master
Mass production management	Parts drawing			Instructions	Notebook

Figure 7.9 Transition of design tool and information transfer media

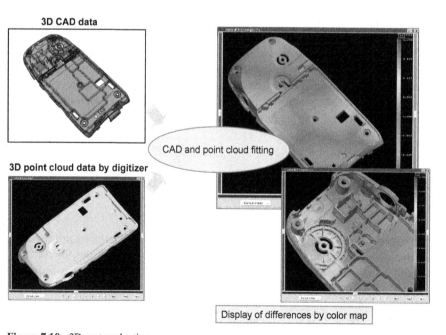

3D CAD data

CAD and point cloud fitting

3D point cloud data by digitizer

Display of differences by color map

Figure 7.10 3D part evaluation

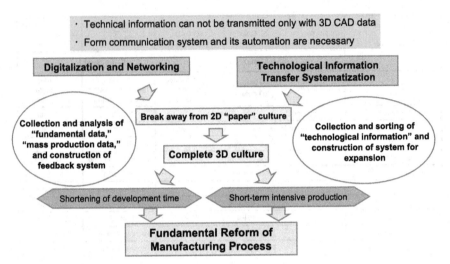

Figure 7.11   Efforts for drastic reforms in manufacturing process

As shown in Figure 7.11, the company was finally able to switch from 2D paper drawings to digital data with the appearance of XVL, the 3D data format which allows 3D and attribute expression, and information integration tools such as XVL Web Master and XVL Notebook. These tools allowed them to share part attribute information among sections, enabling drawing-less communication. As mentioned above, 3D data is unable to convey all the technical information required for manufacturing. What is required is a system which automatically links technical information to 3D and distributes it digitally. With such a system, YAMAGATA CASIO was able to break away from their long practice of having to use 2D paper. By thoroughly implementing the use of 3D for both software and hardware, they succeeded in establishing a 3D culture, shortening development time, and realizing short-term intensive production. This is an epoch-making reformation of the manufacturing process at the company.

7.7 XVL-based Technical Information Distribution Key to Success

There were three keys to YAMAGATA CASIO success in digitalizing information distribution, as shown in Figure 7.12. The first was setting down rules for conveying technical information to people outside the company and between different CAD systems. The company says it was necessary to establish rules to ensure consistent communication of design intent – especially tolerance, finished state, and jointed state. The second key was automating the translation and flow of information. Busy designers tend to shun tedious work, and to them, the job of having

① **Need for creating files for technical information transfer**

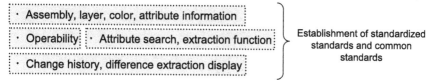

· Between in-house department · Between different CAD · Between companies

② **Need for automation and labor saving in information output**

· Parts list · Machining instructions · Hand chaff cutter side, product side, sliding side, etc.

③ **Need for development of viewer which can be used in downstream processes**

· Assembly, layer, color, attribute information

· Operability · Attribute search, extraction function

· Change history, difference extraction display

Establishment of standardized standards and common standards

Figure 7.12   Further progress of information distribution in future by 3D data

to convey information to the downstream processes is needless work. So, such simple work must be automated as much as possible. The third key was the availability of a 3D viewer which can be used in downstream processes. It was necessary to distribute shape information, as well as assembly, layer, and attributes. It was also important to be able to search for attribute information and extract what was necessary from a massive volume of information. And of course, they had to accurately convey record of changes and differences before and after changes. Currently, the company uses XVL Notebook for this. Now that the company has a fully 3D infrastructure, it will be even more important for the company to enhance the means of expressing and conveying the design and technical information to support their manufacturing work.

Chapter 8
ALPINE PRECISION:
Report-less and Drawing-less in Mold Making

ALPINE PRECISION plays a core role in production activities at the global company ALPINE which manufactures on-vehicle information systems such as car audio and navigation systems. It mainly manufactures on-vehicle equipment mechanical units and exterior components. ALPINE PRECISION also serves as the mother plant of ALPINE's overseas production bases.

ALPINE PRECISION was an early adopter of 3D CAD for their design activities in order to promote 3D design at the company. In 2003, the company moved to completely drawing-less and report-less information exchange, and started using XVL to share inter-department information including offshore departments. The company is seeing great returns from using 3D data throughout their design and manufacturing activities, from mold design to production and maintenance. This chapter introduces the pioneering case study of ALPINE PRECISION as explained by Mr. Shigeki Yoshihara and Mr. Nobuyoshi Mizuno of the Die and Mold Making Department of the company.

8.1 Weapons for Global Expansion and Delivery Time Reduction

Trends such as global expansion of the Japanese car industry and shorter market cycles have brought such tasks as globalization and shorter delivery times to on-vehicle equipment manufacturers. Increasingly short delivery times are compressing the deadlines for die and mold making. Meeting these shorter deadlines requires fast exchange of information between die and mold design departments and manufacturing departments. Sharing 3D shape information and related attribute information not only reduces losses in information transfer, but also enhances productivity. Also, adding mold manufacturing data enables manufacturing information to be shared with overseas departments as well. The use of 3D data is a powerful weapon for global expansion of the manufacturing industry as well as for shortening product delivery time.

ALPINE PRECISION has used 3D CAM systems in their mold-making activities since the 1990s, and started using 3D CAD for design in 2000. They adopted digital manufacturing to increase the quality of their mold making. However, they found that were not able to make full use of 3D CAD data because they had problems sharing 3D information and related attribute information inside the company.

To maximize sharing of digital data – and ultimately eliminate all use of drawings and reports – the company launched a campaign to use lightweight 3D data in 2003.

8.2 Limitations of Business Activities Based on Drawings and Reports

As shown in Figure 8.1, ALPINE PRECISION faced three major problems with mold-making methods based on conventional drawings and reports. First was the need to prepare 2D drawings even though they were doing 3D design. The preparation of drawings in some ways relies on individual skills, *i. e.*, how to prepare the cross-section or how to express size. The parties reading these drawings also need corresponding skills. In particular, various difficulties may be encountered when the exchange is between people from different countries with differing educational backgrounds. Another problem is when the design and manufacturing process is carried out based on 2D drawings, the different departments involved are unable to carry out their respective work concurrently. In order to shorten delivery, work such as acquiring parts information beforehand and starting manufacturing preparations must be done concurrently, and this is difficult to do based on drawings.

The second problem was the need to prepare massive volumes of reports for mold making. Reports need to include all the necessary information for the manufacturing process in a clear-cut manner, and hours are spent just preparing them.

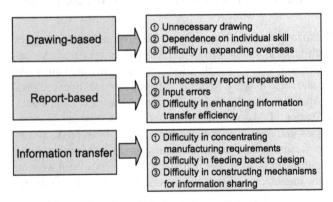

Figure 8.1 Problems in mold design and manufacturing process

When a task involves massive work, data input errors are apt to occur. Naturally, this results in major losses in the downstream process. As long as paper-based jobs such as reports are required, it is difficult to enhance work efficiency and quickly transfer information to other departments.

The third problem was the transfer of information. Usually, work methods based on drawings and reports involve the physical distribution of a range of information on paper to various departments, which makes it very hard to share the information required for mold making as well as to focus on important manufacturing conditions in the manufacturing process. Moreover, this method makes it difficult to send manufacturing feedback and requirements upstream to design. As mold-making bases expand outside the country, it is becoming increasingly important to transfer information from headquarters to the various manufacturing bases and departments spread out all over the country and the world.

ALPINE PRECISION found that they were slowly reaching a dead-end in improving work efficiency in operations based on drawings and reports. The company realized that the method to resolve the entire problem was to incorporate the attribute information required for mold making into lightweight 3D data and share this data throughout the company. They found that in order to be used efficiently, 3D mold data had to be light, have high display accuracy, and have a good response. Even the inexpensive PCs at the manufacturing department had to be able to display large 3D mold models, and the 3D viewers had to be easy to use for manufacturing staff not accustomed to complicated application software. The viewer also had to be able to handle attribute information with the 3D data. Based on all these requirements, ALPINE PRECISION decided to introduce XVL into their business activities/solutions to innovate their mold-making process. The following describes how they have succeeded in doing so using 3D data in their mold-making processes.

8.3 Use of 3D Data for Mold Design Review

When a mold design department designs a mold, they will review the mold from procedural and manufacturing perspectives with staff from related departments. At ALPINE PRECISION, they performed these reviews using 2D drawings derived from the 3D CAD models. However, review by drawings had various problems; those who do not have the skills to read drawings found difficulty in understanding the proper shape, and because some of the staff from departments other than design did not have such skills, they were unable to participate in the review.

In addition, reviews could only be started after the drawings were ready. This meant a delay in reviews if the drawings were late. ALPINE PRECISION therefore started converting mold shapes to XVL for review by their technical and manufacturing departments (Figure 8.2). Since XVL was generated automatically when the design was complete, these departments could review the models promptly and accurately. Reviewers were also able to understand the design intent much better.

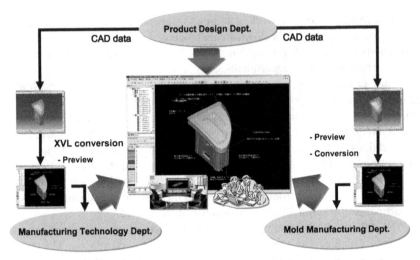

Figure 8.2   Use of 3D data in product review by design/technology/manufacturing departments

Before the company started to use XVL, the design department wrote comments on the CAD models, prepared screen shots (images) of the displayed comments, and gave these to the manufacturing department for review. When the company introduced XVL, they started adding design review (DR) results directly to CAD models as annotations, and then converted this to XVL. This system provides clear, simple communication from the design to the manufacturing department.

In this way, ALPINE PRECISION started having the three departments related to mold making – design, technical, and manufacturing – preview the 3D models from their respective standpoints. After the previews, people from the three departments would meet, go over the preview comments, and carry out the DR using XVL. This preparation greatly improved the efficiency of the DR. XVL itself serves as an excellent easy-to-understand visual record because comments are written down on the 3D models themselves. Also, lightweight, digital XVL promotes paperless communication as well as information sharing at the corporate level. Subsequently, this enables manufacturing requirements to be fed back to the upstream design department, enhancing the design quality.

8.4   Company-wide Sharing of Design Information

In order to be able to use 3D XVL mold data throughout the company, the XVL data must contain more than shape information. It must also include production specifications and parts information, data that is important to downstream processes. Designers can add this data to the 3D CAD model, and it will automatically be converted to XVL. Being light, XVL is capable of handling shape information

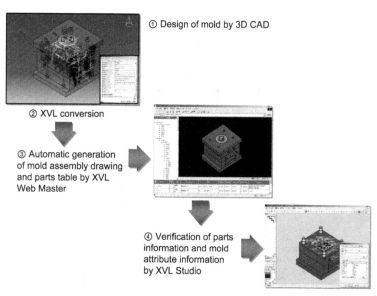

① Design of mold by 3D CAD

② XVL conversion

③ Automatic generation
of mold assembly drawing
and parts table by XVL
Web Master

④ Verification of parts
information and mold
attribute information
by XVL Studio

Figure 8.3 Development of mold information at mold design department

and attribute information on the web at the same time. The tool to support this is "XVL Web Master." As shown in Figure 8.3, mold assembly drawings and parts list can be automatically created from XVL and shared on the network. Parts information and mold attribute information accompanying XVL can also be browsed using XVL Studio, the viewer software for XVL. This XVL Studio software has a measurement function which can measure shape size accurately as well as display cross-sections. This means that as soon as the design department creates 3D and attribute information, this information can be shared with the manufacturing department. At ALPINE PRECISION, their manufacturing department is now able to obtain accurate information instantaneously from the web in the XVL format (Figure 8.4). Manufacturing processes are now completely drawing-less since the data for all steps in the manufacturing process can be downloaded from the web. This in turn has also eliminated reports from the manufacturing process. Since information such as manufacturing size and conditions can be added to XVL and shared by the whole company, mold quality is improving. The manufacturing department then adds manufacturing data such as sizes to the XVL files. Assembly information is also added in the mold assembly stage, completing the required 3D shape and attribute information. The final XVL data, including all the manufacturing information, is shared by the whole company.

3D communication has many advantages over drawing-based communication. For one, it enables the correlation between parts to be understood intuitively, allowing the manufacturing department to carry out their work more precisely and efficiently. Those not familiar with drawings at the manufacturing site are also able to see the designed shape in 3D, so information which would be vague on

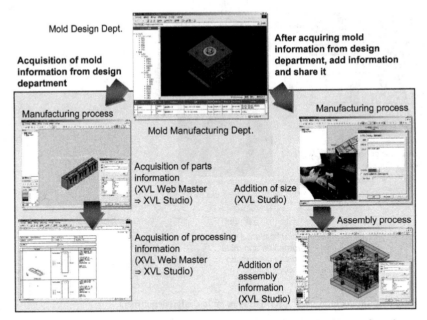

Figure 8.4 Acquisition of mold information and addition of information in manufacturing process

drawings is much clearer in 3D. Also the data can be measured and cross-sections checked visually. This increased understanding reduces the number of questions for designers and increases work efficiency throughout the company.

In this way, ALPINE PRECISION succeeded in switching from a drawing-based and report-based communication system between design and manufacturing

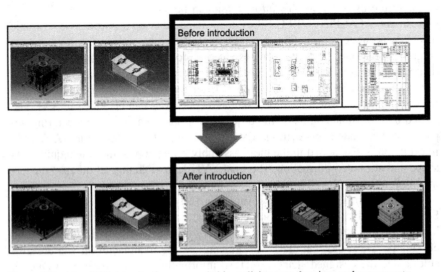

Figure 8.5 After XVL introduction, succeeded in realizing zero drawings and zero reports

to an XVL-based communication system. This has increased design efficiency by 40%. Back when they were using drawings, they also needed the time to decide whether to draw the cross-section on paper and where to set the tolerance. With the new method, there is no longer the need for such drawing skills, resulting in zero human drawing and report errors.

As shown in Figure 8.5, drawing information is replaced by XVL and reports by XVL generated by XVL Web Master and attribute information. Compared to 2D drawings, XVL provides significantly more information volume. Using the time saved to perform more DRs, improves the design quality even more. The information flowing between the mold design department and manufacturing department is accurate and prompt, and this helps deepen the relationship between the departments. Enhanced communication between departments contributes to enhanced productivity both tangibly and intangibly.

8.5 Review by Mold Manufacturing Department

The manufacturing department at ALPINE PRECISION carries out simple operational simulation using XVL to review problems before trial production (Figure 8.6). At this time, manufacturing conditions such as actual mold size are determined. The review results are added to the file as mold history information using XVL Studio. Publishing the XVL data to the web using XVL Web Master makes it easy to share. In addition to sharing the information with downstream departments

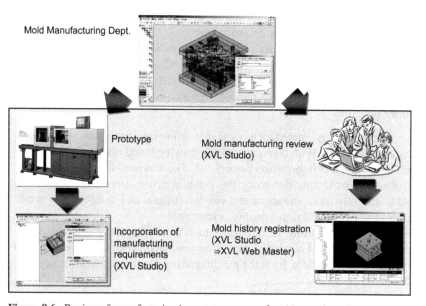

Figure 8.6 Review of manufacturing in prototype stage of mold manufacturing

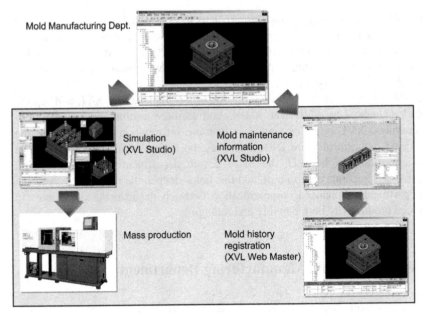

Figure 8.7   Review of manufacturing in mass production stage of mold manufacturing

like maintenance, the manufacturing information is also fed back to the upstream design department. In this way, ALPINE PRECISION also uses XVL for previews before mass production (Figure 8.7). Defect information in mass production is registered as maintenance information and converted to a form which can be shared on the web. This information can be used as information for the development of new molds and maintenance procedures.

Another benefit of drawing-less and report-less mold making is it allows manufacturing problems to be promptly broadcast to related departments like design and maintenance. Such activities enhance mold finish quality and manufacturing speed. Overall, the use of XVL in mold design and manufacturing processes provides three benefits:

1. Parts information is immediately available, allowing manufacturing preparations to start earlier. With conventional 2D drawing based operations it was difficult to obtain parts information beforehand. This has been improved.
2. All related departments can share the manufacturing information, enhancing mold quality. By incorporating actual size information and mold-making conditions in XVL, the mold finish quality is improved.
3. By adding manufacturing requirements to XVL during the mold-making review, manufacturing knowledge can be accumulated. This allows feedback of manufacturing conditions from the manufacturing department to the design department.

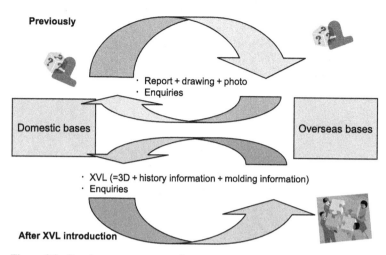

Figure 8.8 Development to overseas bases

Such information sharing can also be implemented overseas (Figure 8.8). Until now, when a problem developed abroad, the defect was photographed and discussions were carried out based on these photos. This method of specifying a problem with offshore bases using countless drawings and reports was slow and cumbersome. It made it difficult to establish a prompt troubleshooting system. Communication using 3D data ensures clear and prompt understanding. Lightweight XVL is therefore useful for overseas information sharing. By sharing mold history information and parts information with production bases abroad, problems occurring outside the country can now be responded to and handled easily. Use of XVL as a communication tool provides the required information to the department needing it, accurately and immediately.

8.6 Application of 3D Data to Manuals

ALPINE PRECISION plans to centrally manage XVL data in a database system. They also plan to implement a security system that will enable them to expand XVL information sharing with the overseas division and partner companies.

The company is also planning to use XVL animation and illustration functions in preparing manuals. Armed with the powerful "weapon" that is information, their mold design and making process will become a solid fortress for their manufacturing activities.

Chapter 9
TOKAI RIKA:
Visualization of Manufacturing Information Mold Making Using 3D Work Specifications

TOKAI RIKA is a manufacturer using 3D data in their operations. The company's core business lies in the production of car parts. To enhance their in-house mold-making operations, the company decided to switch from paper to 3D data. The company proved very successful by using a unique method of coloring work specifications. They are using 3D to improve engineering communication in mold design.

9.1 Tasks and Solutions in Mold-making Departments at TOKAI RIKA

The increasing efforts made to shorten the product lifecycle of cars are also pressing parts manufacturers to reduce their delivery deadlines as well. This is making it essential for them to revise design and manufacturing processes and enhance work efficiency. In order to improve efficiency, TOKAI RIKA started using 3D CAD for mold design. However, their manufacturing department continued to use 2D drawings. To improve the mold-manufacturing process, TOKAI RIKA started using XVL in 2003. In the old process, when a mold design was completed the designer would prepare drawings that were colored to specify the required machining accuracy and work instructions (Figure 9.1). However, it was often difficult for the engineers in the manufacturing department to accurately grasp the intended shape from drawings alone. They then had to check with the designers. This limited manufacturing efficiency, and both the engineers and designers had to deal with needless work. So, the company decided to use 3D data for their mold-making operations. They looked for an easy-to-use 3D viewer for their departments. They wanted a viewer where: (1) the surface color could be easily changed, (2) it is easy to measure, and (3) they are able to convert from various kinds of

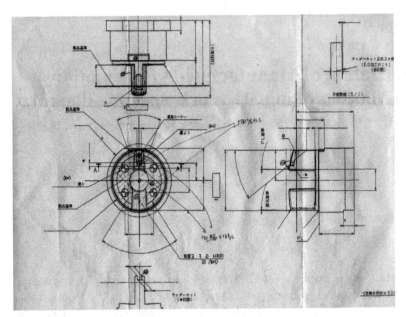

Figure 9.1 Conventional mold drawing (drawings were colored)

3D CAD data to viewer data. There were several viewers available meeting these requirements, so they further added three more requirements: (4) lightweight data, (5) able to run on existing low-end PCs, and (6) could be supported by TOKAI RIKA's information system department. Based on all these requirements, the company chose Lattice XVL.

9.2 Using XVL and Advantages

In the past, after designing the mold shape on the 3D CAD system, the design staff would prepare drawings for the manufacturing engineers and send them to them upon completion. This meant the work was sequential, not concurrent. As shown in Figure 9.2, in the XVL-based process, when 3D design is completed the CAD data is converted to XVL and sent to the mold-making department. The manufacturing engineers then add the machining accuracy and work process to the XVL data.

Of course, the design department would have already specified machining accuracy and machining method using different colors on the CAD model, and this color information is maintained in the XVL model. The manufacturing department would use this information and, taking into account the operational state of the machining facilities, would decide which machine to use for each surface: die-sinking electric discharge machining (EDM), wire EDM, or general machining. They would then change the color of each surface accordingly (Figure 9.3).

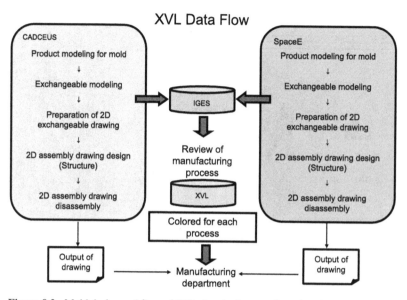

Figure 9.2   Mold design and flow of XVL data in the manufacturing department

The company's use of XVL Notebook to prepare NC programming specifications is also unique. XVL Notebook is a tool for preparing dynamic 3D documents. It is able to link 3D shapes and images and show the whole mold and detailed drawings of the mold. It is similar to the detailed and overview maps displayed on car navigation systems. The NC programming experts at TOKAI

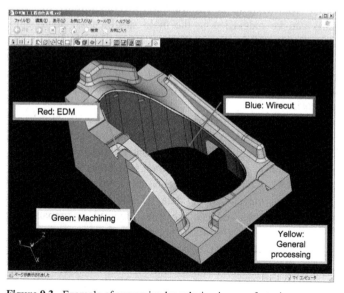

Figure 9.3   Example of expression by coloring in manufacturing process

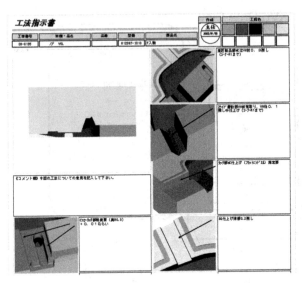

Figure 9.4 NC programming specifications from XVL Notebook

RIKA, whom they call "mold meisters," use this tool to give program instructions and pass down their knowledge to the other staff. For example, they would be able to give the instruction "this is to be machined in the general process later, so leave a 1-mm machining allowance" in detail on XVL Notebook (Figure 9.4). This tool shows visually the parts for which the instructions are given.

The NC program specifications are then placed on a common server, so that they can be referred to by the required staff. Staff members preparing NC programs on CAM systems follow these instructions when programming the machines. Machined parts can be magnified in the mold specification documents created on XVL Notebook. Since the magnified views and overall 3D display are linked, users can tell at a glance which part of the whole mold they are looking at. With conventional drawings it was not always clear which parts were being shown. XVL Notebook has enabled mold meisters to share their experience and knowledge clearly and easily.

The mold-making department took a strong initiative to use XVL Notebook documents for other purposes as well, such as the measurement instructions for the part shown in Figure 9.5. Measurement specifications indicated by overview drawings take a very long time to prepare, and at the same time, they are often difficult to understand. In XVL Notebook documents, the measurement location can be checked by linking the images and 3D data. This dramatically enhances understanding. Also, using 3D data reduces the time required to prepare the specifications. So, by providing 3D XVL data, the design department helps to increase the efficiency of manufacturing.

Figure 9.5 Mold measurement specifications from XVL Notebook

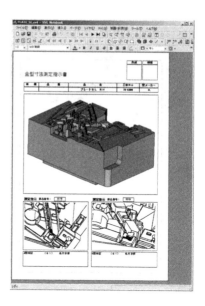

9.3 Using XVL in the Manufacturing Department

Some employees at the design department opposed the company's plans to use XVL for their design and manufacturing processes. It is only natural for any workplace to find it hard to change from a familiar method. When the company first started using XVL, the mold design workload increased. For example, they now needed to add colors to the models. However, the company was able to persuade the department to cooperate by convincing them that this extra work would dramatically improve downstream processes. At the same time, there were also strong supporters of XVL in the mold-manufacturing department. They actively involved to introduce the new method and started using it to prepare operational manuals. With the company-wide mission to reduce mold design and manufacturing lead time, the mold-manufacturing department started to use XVL aggressively. It was accepted that drawing-based processes could not reduce the lead times, so the department adopted XVL to do so. Figure 9.6 shows how the manufacturing department uses XVL to verify shapes.

In 2002, the information systems department started using educational presentations to promote the use of XVL throughout the company. XVL educational presentations were carried out in sections and in groups. Various materials were prepared for the presentations, from technical introductions to case studies, and the presentation was tailored to each department. Key manufacturing personnel were consulted for the case studies. As part of the effort to reduce lead times, management welcomed and encouraged these efforts. Sometimes, the management themselves would also promote the use of XVL together with the information system department. Such efforts paid off, and gradually XVL started to spread within the company.

Figure 9.6   Checking XVL at worksite

9.4 How 3D Has Improved Operations at TOKAI RIKA

By using XVL, the mold design department no longer needed to specify detailed dimensions. Unfortunately, some of the partner companies of TOKAI RIKA are still using drawings in their work, so the company has not achieved a complete drawing-less environment, but they have certainly been able to simplify drawings a great deal. When someone working on the machining process needs to know the size of the mold, *etc.*, all they need to do is to directly measure the XVL data. Before, whenever size information was missing from drawings, the manufacturing department would blame the design department.

Another benefit of XVL is better manufacturing time estimation. 3D data provides more accurate measurement of the machining scope, machined area and perimeter, resulting in better overall time estimates. Machining ends at the expected time, allowing mold making to be carried out as planned. Another enormous benefit is that XVL is inexpensive to deploy. For example, viewing XVL requires only an inexpensive PC. Editing XVL and measuring XVL models require only low-cost tools.

The unique aspect of TOKAI RIKA's XVL implementation is their method of expressing the required machining accuracy by coloring the model surface. In the past, the manufacturing department added colors to drawings, but this work was no longer required once the company started using XVL. Eliminating this task provided a dramatic reduction in man-hours. With drawings, they also had a problem with complicated shapes because they were unable to paint hidden parts. With

3D shapes, color can be added to any surface. In addition, the manufacturing department has its own standardized coloring rules which allow anyone inside the department to understand the machining details by looking at the colored model. The mold design department was using different CAD systems, but now they use XVL as the common format for their manufacturing process. By introducing XVL, even the mold-making department is beginning to appreciate the advantages of 3D, in other words, the staff at their manufacturing plants are also starting to experience the same benefits experienced by designers when they switched from 2D CAD to 3D CAD.

9.5 From 2D Drawings to 3D Drawings

Gradually, TOKAI RIKA started to expand the use of 3D tools from molds to product design as well, and interest in XVL continues to grow. They are currently constructing a system for generating XVL for every 3D CAD model. In product development, when the initial design is complete, related staff would be notified via email and would meet to review the design. Since staff outside the design department did not have access to CAD systems, XVL would be used to review the design. This is one example of how, at TOKAI RIKA, XVL is becoming the standard 3D format. XVL can be pasted into documents, and downstream departments – production, inspection, and manufacturing – are starting to use it too. These departments are accustomed to using drawings in their work, and have been reluctant to switch to 3D tools. However, as 3D design started to diffuse through the company, interested started to pick up. Subsequently the company required that all drawings be 3D, further accelerating the use of XVL.

XVL has been well received as an assembly process design review tool (Figure 9.7). So TOKAI RIKA constructed a system based on XVL Web Master to send 3D product assembly instructions to their manufacturing plants and overseas manufacturing bases. They are finding 3D to be especially useful for overseas communications where language is a barrier. The company is also encouraging their partners to use XVL. Very few of these manufacturers have 3D CAD, and when faced with the need to use 3D data, they find XVL to be very useful. TOKAI RIKA provides 3D CAD data to mold-manufacturing subcontractors who use 3D CAD, but XVL data has been more than sufficient for subcontractors who do not use CAD.

When a company needs to release 3D data to external parties, security is an issue. This means that integrated security measures are important for documents with 3D data. At present, TOKAI RIKA exchanges XVL data according to the security rules for 3D CAD data exchange. The company is also planning to distribute 3D shape information and product assembly animations to Japan and overseas mold-maintenance bases. They plan to establish practical security rules to expand the scope of 3D data use. TOKAI RIKA is a typical example of a manufacturer promoting the use of 3D data by replacing color drawings with XVL. By enhancing

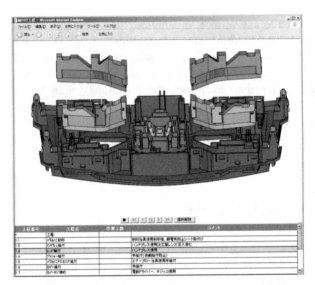

Figure 9.7 Assembly process distribution from XVL Web Master

their business processes with 3D data they have dramatically improved their efficiency. TOKAI RIKA is also an interesting case where the use of 3D data by the manufacturing department encouraged use at the design department. This case study shows how lightweight 3D is surpassing 3D CAD/CAM/CAE in use.

Chapter 10
CASIO: Creating Customer Manuals Using 3D Data

CASIO, a leading manufacturer of digital cameras, electronic dictionaries, and other digital equipment, is a pioneering case study of 3D documentation. Technically creating illustrations from 3D data is easy; what is difficult is deciding who is to prepare the 3D data and how, and who is to encourage the use of this 3D data. This chapter discusses the efforts of Mr. Mitsuhiko Iwata, Manager of the Development Department, Technical Division, Design System Development Group, and Mr. Hideo Kashiwaguma, Manager of R&D, CASIO. XVL was first applied to user documentation and then it spread throughout the company as a communication tool.

10.1 After 3D Design Practice Started Kicking In

CASIO develops and manufactures a diverse range of electronic equipment, from consumer goods, such as calculators, electronic dictionaries, digital cameras, electronic instruments, and watches, to information equipment. The product lifespan of electronic equipment is growing shorter and shorter, and this is driving shorter and shorter product development cycles. The Development Department at CASIO develops and supports design IT tools for all products that the company manufactures, and it is at the lead of shortening development cycles. Figure 10.1 shows the timeline for the introduction of CAD into the company's business activities. They started to use 3D CAD in the early 1990s, and in 1998, they chose PTC "Pro/ENGINEER" as the company's standard 3D design tool for PCs. They actually started using 3D data only from 2002 after they had accumulated an adequate pool of data. They applied 3D data to a broad range of business activities and dramatically improved their manufacturing process. As shown in Figure 10.2, CASIO's goal in using 3D data was to apply in their design, quality, mold, and materials departments to restructure their business. They called this "Single Source Multi Use." CASIO chose XVL because it is the lightest and the easiest to use 3D format.

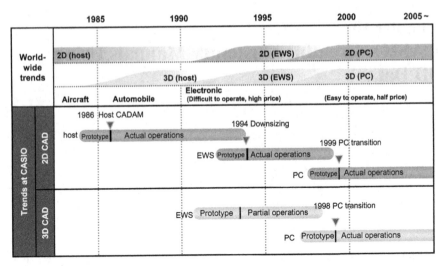

Figure 10.1 Adoption of CAD at CASIO

Figure 10.3 compares the data volume and data transfer time between XVL and CAD data. For instance with the CASIO EXILIM digital camera they are able to compress 20 MB data down to 0.4 MB. At the same time, they are able to reduce the time to send the data to a subcontractor to 1/90 the original time. Because CASIO had been outsourcing manual creation to subcontractors and were plan-

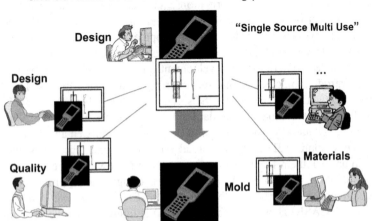

Figure 10.2 Reason for using XVL

• Data Compression Example (EXILIM)

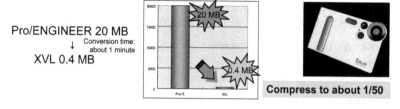

Pro/ENGINEER 20 MB
↓ Conversion time:
about 1 minute
XVL 0.4 MB

Compress to about 1/50

Reference: Transfer time to site outside company by FTP

Pro/ENGINEER	XVL
931.1 s	14.5 s
842.4 s	9.4 s
666.9 s	5.4 s
635.1 s	6.7 s
889.7 s	10.2 s
Ave 793.0 s	Ave 9.2 s

※ Pro/ENGINEER generated data
is zipped and sent

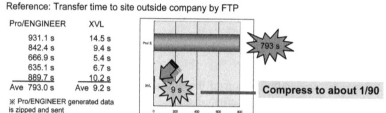

Compress to about 1/90

Figure 10.3 Data compression and transfer evaluation using actual data

ning to send 3D data to the subcontractors for use to prepare illustrations, nothing excited them more than the fact that they were going to be able to shorten the data transfer time.

10.2 e-Manual Project

CASIO first used XVL to create owner manuals and service manuals for their products. They called such manuals e-manuals. The goal of the e-Manual Project was to start the manual creation work early using CAD models currently being designed to generate the manual illustrations. They wanted to use the CAD models to frontload the manual production process and shorten the manual development time. This process also enabled CASIO to estimate the approximate time of manual completion from the design stage. Of course designs could be changed later by notifying their subcontractors of the required changes. To prepare user manuals, CASIO would allocate work to the respective subcontractors, give them most of the instructions on how to prepare the manuals, and have the subcontractors prepare the manuals accordingly. Service manuals, on the other hand, are internal documents and are prepared mainly by subcontractors. The subcontractors create exploded views using XVL Studio. The exploded views are then output as vector data, and read into Adobe "Illustrator" to be finished. In the beginning, CASIO provide CAD data instead of XVL data to the subcontractors. However, the CAD data was large, took a long time to transfer, and was difficult for the subcontractors to use to create images. With XVL they were able to reduce the data transfer time and the workload.

The advantages of adopting e-manuals were immeasurable. First, they saved tens of thousands of dollars annually by reducing the time required to draw illustrations. More importantly, however, were the manufacturing benefits that:

1. Achieved faster delivery time by starting manual preparation early using the design data
2. Eliminated errors by giving the manual creators product shape and structural information
3. Reduced enquiries to the design department during the manual creation process
4. Eliminated the need for the sample products that were necessary for manual creation in the past
5. Completely eliminated inefficient large data transfers for musical instruments and printers

CASIO designers especially welcomed the dramatic decrease in enquiries regarding sample assembly and manual creation during busy periods of development.

The e-Manual Project was launched in 2002 and covered 12 models – from digital cameras to musical instruments – in the first year. After proving its advantages, it was applied to many more products. Readers of the manuals were also pleased because illustration quality was more consistent and there were fewer errors. The digital documents are enclosed with products as well as distributed on websites. The service manuals are distributed to several thousand recipients on CDs or via the Intranet. These documents use the vector data generated from XVL.

Figure 10.4 shows manuals made using XVL. These are the actual instructions manuals of the CASIO EXILIM digital camera and XJ Projector. The company is now also using XVL for internal technical documents. They use image data generated from XVL for preparing 3D assembly drawings and quality assurance reports.

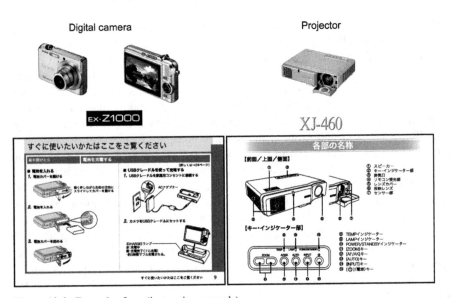

Figure 10.4 Example of use (instruction manuals)

Figure 10.5 Example
of use (service manual)

3D Service Manual
(Example of digital camera EXILIM)
- Created at service dept. for limited time
  web release for overseas dealers

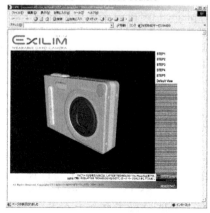

In the case of these internal documents, 2D images are sufficient. However, with the use of 3D data in e-manuals becoming a standard practice, CASIO is now attempting to use 3D data for communication. As shown in Figure 10.5, CASIO has been creating 3D service manuals for its overseas dealers, and they have found 3D animation to be excellent for these manuals. Animated service manuals are created by the service department and distributed on the web. Descriptions using 3D animations are much easier to understand than text descriptions. CASIO has also started to use viewer products to employ XVL as a communication tool (Figure 10.6). In this way, the use of XVL has evolved starting with user documentation, then animated service manuals, and finally as a general 3D communications tool.

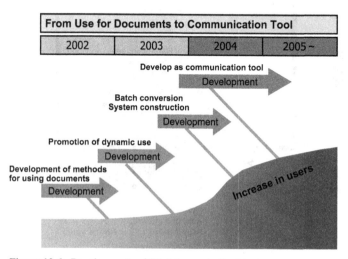

Figure 10.6 Developments of 3D data uses in the future

10.3 Driving Force Behind Use of 3D Data

So what is the driving force that helped CASIO use 3D data smoothly? This can be explained by three key points: collective conversion to XVL (batch conversion), expansion support activities, and support from management of the company. First, what is batch conversion to XVL? To promote the use of 3D data, people requiring it must have access to it. This is made possible by a web-based conversion system. CASIO uses Pro/INTRALINK to manage their 3D data (Figure 10.7). This system also provides a web interface. Anyone can log in and request a model, and the system will automatically convert it to XVL and send it to them. There are two main benefits to this system. The first is the designer never needs to deal with additional work. All the designer needs to do is to check the CAD data into the data management tool, after that anyone can convert the data anytime they want. The second is that anyone requiring XVL data is able to acquire it very easily just by accessing the web. CASIO spent considerable effort developing the batch conversion system, developing one that is friendly to both the designer and the downstream user. CASIO now also uses the system with its subcontracted manual creators.

The second driving force behind the use of XVL was the promotional activities and materials shown in Figure 10.8. These activities and materials were created by the development head office and included:

1. Introductory seminars about XVL
2. Training sessions for XVL tools
3. In-house web portals containing XVL support materials
4. Simple XVL tool manuals

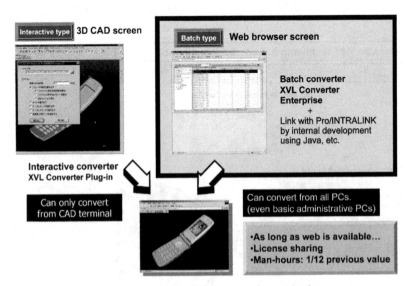

Figure 10.7 Construction of web-based XVL conversion mechanism

XVL Development Activities
- Introductory seminars
- Training sessions
- In-house portals
- Simplified manuals

Figure 10.8 XVL development support activities

This dedicated effort helped spread XVL by showing users the advantages of 3D data. Users who learned the benefits of 3D data from other departments also started to use it in their own activities.

The third driving force behind the use of XVL was the understanding and support of company management. When XVL was first adopted and shown to be effective at improving business processes, CASIO's executives strongly backed its use. The management of the company welcomed the best practice stories of XVL and saw it was an outstanding tool that would enhance the business. Such strong support was bound to spread the use of 3D data throughout the company. In 2003, the group promoting the use of XVL was awarded the president's prize. Such high recognition motivated everyone throughout the company to start using 3D data.

10.4 Online Data Reviews

CASIO's use of XVL rapidly expanded from documentation to a communication tool and is now being used throughout the company (Figure 10.9). The service department creates animated 3D service manuals and the manual department uses XVL to create user manuals. The design department uses XVL for meetings with manufacturers while the engineering department uses it for drawing inspection and

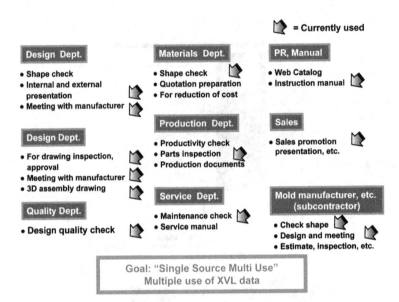

Figure 10.9   Development state by department

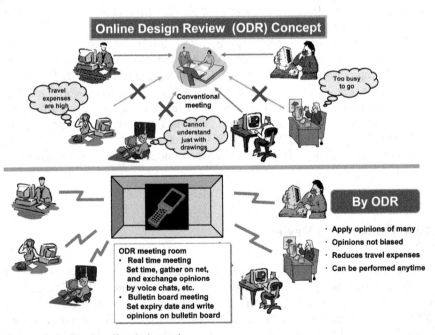

Figure 10.10   Use in online design review

approval. CASIO also uses XVL for shape verifications during meetings with mold manufacturers. With use of XVL growing in the design, engineering, manual, and service departments, the company plans to further expand use in the materials and production departments. Currently, XVL is used company-wide for online design reviews (ODR). Previous design reviews required bringing all related parties together in the same place to review engineering drawings. This was expensive and parties often could not attend because they were too busy or were unable to understand the engineering drawings. The ODR resolves these issues. As shown in Figure 10.10, the ODR is held online so anyone can participate from anywhere. The participants communicate by phone and online bulletin board. Also, the ODR uses XVL models that are much easier to understand than engineering drawings. There are no time constraints, and everyone is free to participate in the design review. In the future CASIO plans to use XVL in more business activities. With digital equipment lifespan becoming increasingly short, the product development competition is intensifying globally. We can indeed look forward to the future developments at CASIO, a company one step ahead of others in the use of 3D data.

Chapter 11
KVAL: 3D Information Sharing and Its Effects at a Middle-scale Firm

Is the use of 3D data only the luxury of big companies? No, there are ways for small and medium-sized companies to enhance work efficiency using 3D data, as proven by KVAL Inc., a manufacturer of woodworking machinery in the USA with an employee body of 100. Says Sebastien Jame, Engineering Services Director at KVAL, "Within just a few months of adopting XVL compression technology we have already seen major benefits to our business. Conservatively speaking productivity gains of 20–25% in manufacturing, technician learning curves reduced to minutes instead of days, and spare part service calls often reduced from 15 to 20 minutes to just seconds." This chapter introduces the use of 3D data at a middle-scale firm as seen from the case study of KVAL.

11.1 Use of 3D Data for Maintenance of Complicated Machines

KVAL is a family-owned company that has been designing and manufacturing heavy industrial woodworking machinery since 1947. For 60 years, KVAL has been providing high-quality machinery, parts, and service to the millwork industry and is especially recognized as an industry leader in the design and manufacture of precision-engineered, solidly built door machining equipment. Increased competitiveness within the millwork industry has forced many smaller companies to become more specialized. KVAL has successfully focused on customizing the design and configuration of many of its machines to the needs of specialized customers. Today, KVAL creates, ships, and installs over 300 unique machines/year. As shown in Figure 12.1, the machines made by KAL are very complicated, and often a single machine can have up to 6,000 individual parts. The company already maintains more than 10,000 machines in the field.

Founder A.A. Kvalheim started the company by designing and building a panel saw employing an ingenious traveling carriage, and has fostered the company to the forefront of design and manufacturing of precision-engineered door machining

Figure 12.1 Comprised of KVAL computer-controlled manufacturing machines and many parts

and millwork equipment. In recent years KVAL adopted SolidWorks, a powerful 3D modeling solution, to quickly and accurately provide the necessary custom design services. They also tried to use SolidWorks' eDrawings capability to share 2D and 3D designs downstream of engineering to find ways to improve their business process. They soon found an appropriate tool for 2D data, but things were not as smooth for 3D data, because the size of the 3D files for KVAL machines is massive, and the consequent time to load, view (zoom/pan), and manipulate the models was too slow with normal tools.

XVL helped resolve this problem. They did an evaluation and found that they could easily create 3D files at a fraction of the size of the eDrawings files and view them with high performance even on low price PCs. The combination of SolidWorks to create 3D models and XVL to share them has decreased assembly and reworking time by as much as 20%, enabling them to enjoy the benefits described at the start of this chapter. Even for middle-scale companies of the size of KVAL, there exists "heavy doors" between the different sections within the company. KVAL was able to open these doors by using a new communication method based on 3D data.

11.2 Opening the Door Between Design and Manufacturing

To a specific customer's requirements, Engineering would design a variant automatic door milling machine and send it "over the wall" to the company's vast production floor, and then not hear anything back for a while. They assumed everything was okay. But often personnel would find difficulties in the design's manufacturability and send a set of marked-up paper drawings back "over the wall" to engineering. The two departments communicated in this manner, not just because of their physical separation, but because the "wall" consisted mainly of a technological barrier. Design spoke CAD; Manufacturing talked red ink on paper.

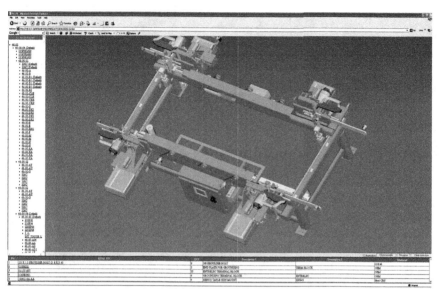

Figure 12.2   KVAL machine XVL model in manufacturing

After installing Lattice 3D technology in the Design department, which ultra-compresses the CAD models, KVAL soon found the benefits. Figure 12.2 shows an example of the design data. In order to convey the mechanism of multiple machines accurately, the common communication method in the form of 3D data is indispensable. For example, 3D even helps the welders see more quickly how to arrange and weld the parts in a more efficient sequence. "Overall the Design and Manufacturing workload is reduced because there are not as many CAD changes coming to Design from Manufacturing. On the other side, Manufacturing is now giving valuable input into the Design process. The new 3D XVL technology did not make lobbing projects back and forth over the wall more efficient or faster; instead, it opened a door in the wall. Now the process of design and manufacturing has become a team effort between the two groups: advice from the production floor on the easiest way to build an assembly is now immediately accessible by the designers. Manufacturing had no CAD knowledge, training, or systems but once they obtained the 3D models in XVL format and could use the simple XVL animation tools and free viewers all these benefits started to accrue and the result was increased productivity and reduced rework."

11.3 Use of 3D Data Between Manufacturing and Technical Support

In just a few months, the compression innovation has had a similar effect on KVAL Technical Support and Service. Often the customer, while looking at their

Figure 12.3   KVAL Technical Service department using Lattice 3D models to handle customer queries

installed KVAL machine, is calling from a cell phone and trying to order a spare part. The KVAL technical Support and Service staff used to have to find the right paper drawing, including the variants for that customer's particular machine, while the customer waited for them on the phone. This was a serious problem for a company whose strength lay in providing machines customized to each customer. So they were faced with the task of enhancing service quality. Now they use the XVL Web Master tool for managing parts list and parts shape information. The technical support and service staff themselves display the assembly structure and animated process on their own desktop PCs (the system automatically creates web pages including 3D models) and instantly click on the 3D model of that KVAL machine (Figure 12.3). As the customer describes the problem the technicians zoom and pan to "virtually" see what the customer sees. Delays are avoided, conversations are much shorter, and correct parts are identified by part number. Figure 12.4 shows a 3D model displayed at the Technical Service department. Even when the right drawing is found, it takes a long time to find the parts desired by the customer from complicated machinery. So what often used to take them 15–20 minutes was done in seconds, no mistakes are made shipping the wrong part, resulting in at least 10% higher productivity, savings by selling and shipping the right parts, and more satisfied customers. 3D models built are the Design department in now effectively used by the Technical Service department.

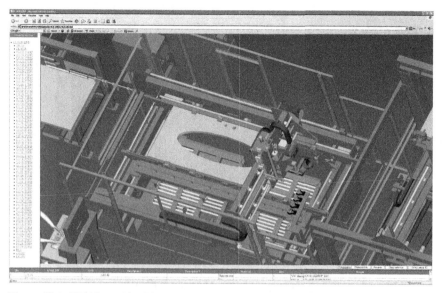

Figure 12.4   KVAL machine XVL model in Technical Service department

11.4 Future Plans: Aiming at 100% 3D

"We are planning to use 3D models for improving communication with customers" (Figure 12.5). Says Jame "Our goal is 100% 3D and we want to give our customers, via the KVAL website or an online WebEx session, the ability to directly view and access a 3D model of their unique machine in order to enable identifying problems and directly ordering spare parts. The customers we've already shown this to are very excited about it." Jame further adds, "We plan to deliver KVAL machines, not with a paper user manual, but with a portable computer loaded with Lattice 3D viewer, the 3D XVL files (including disassembly and assembly animations) of the customers machine, and WebEx so that customers and KVAL Field Service staff can readily understand and work on problems together. With XVL Web Master, a series of components can be shown being put together or taken apart in the correct order of assembly. We are planning to make full use of this function. XVL models and animation provide a powerful tool for technical and maintenance training manuals. What's more, unlike movie or AVI animations, XVL animations can be viewed not just from the camera position but from anywhere. If you want to see how it comes apart from underneath use your mouse to swing it around while the animation is playing! This allows the customer to check machine assembly from the customer's standpoint. This will no doubt sharply reduce maintenance time."

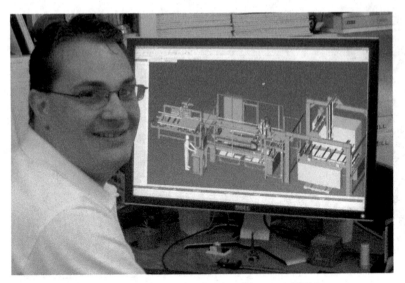

Figure 12.5   Sebastien Jame, Engineering Services Director at KVAL

In time the lightweight XVL models of KVAL machines will be linked by HTML to other information in the ERP system, such as price information, and animated assembly models will incorporate the experience of KVAL assembly specialists so that it can be equally useful for field repairs. "We have really only just started with what Lattice 3D's applications and XVL format can do for KVAL – yet what it has already done for us in just five months has really exceeded my expectations. XVL has been extremely helpful throughout our entire operation," concluded Jame. In this way, at companies like KVAL whose forte lies in providing customized systems to their customers, the use of XVL for communication between departments and with customers should enable them to further increase their competitive strength. KVAL is indeed proof that 3D data can be a powerful source of competitive strength.

Chapter 12
MAN Nutzfahrzeuge AG: Promoting Company-wide Process Chain Using 3D Drawings

MAN Nutzfahrzeuge AG, a leading manufacturer of trucks and buses in Europe, uses XVL as the company's core format, especially encouraging the use of XVL for 3D drawings. Since January 2006, MAN Commercial Vehicles Group has been automatically converting selected 3D design data to XVL and using it for the distribution of 3D design data in the process chain. Currently, the 3D XVL Player is installed on nearly every PC. This chapter discusses the use of 3D data at the company-wide level at MAN based on an interview with Dieter Ziethen, Project Leader in the Information Technology CAE Group, MAN Nutzfahrzeuge AG.

12.1 Using 3D Data for Design, but 2D Drawings for Communication

The MAN Group is a leading manufacturer of vehicles ranging from trucks to buses, diesel engines, and turbomachines, with annual sales of 13 billion Euros and an employee body of 50,000 worldwide. MAN's core areas hold leading positions in their respective markets. As one of the 30 leading German companies, MAN AG (Munich) is a member of the German stock index, or DAX.

MAN Nutzfahrzeuge AG is the largest company in the MAN Group and produces 68,200 trucks and 6,000 coaches per year, has revenues of 7.4 billion Euros, and has 33,000 employees. The company is currently promoting the use of 3D data and has recently adopted XVL.

The MAN Group has a reputation in Europe for constantly integrating solutions to achieve business objectives and ROI. Based on this principle, the company has promoted the use of 3D CAD systems in the company, however information exchange with downstream processes, both within the company and with its suppliers, has been via 2D drawing exchange. Says Dieter Ziethen, "Despite the implementation of 3D CAD systems in design departments, nearly all information distributed remained in 2D. However, it was always our goal that 3D design data, including design information and markups, be deployed across the whole company. The introduction of XVL has enabled us to achieve our goal of sharing 3D information throughout the process."

12.2 Aiming at 3D Communication Throughout the Whole Process Chain

As shown in Figure 13.1, MAN Nutzfahrzeuge's motivation was to improve data quality and enable 3D data for the whole process chain, specifically for the following three reasons:

1. 3D data can be visualized in the whole process chain because it is less abstract than a 2D conventional drawing.
2. Because design is already done in 3D, the design departments will see a reduction in time by avoiding the 2D drawing production step.
3. Non-CAD-trained staff will be able to visualize and interrogate design models thereby enabling easier cooperation on a common data basis and including refinements such as 3D animations of assembly and disassembly and "exploded" part views.

To achieve these goals, the company needed a common basis for communication that would address the whole process, for the whole company, no matter what different CAD systems were used. They therefore adopted XVL for the lightweight 3D data.

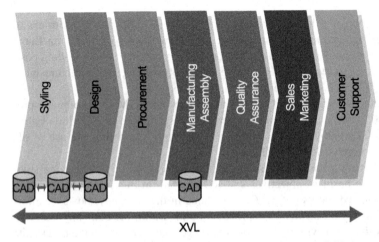

Figure 13.1 MAN process chain: XVL increases opportunities to use CAD data

12.3 Selecting XVL for its Lightweight and Interactive Features

The following are 11 reasons why MAN Nutzfahrzeuge decided to adopt XVL:

1. XVL is CAD-system independent and has converters for all significant CAD systems.
2. XVL is a lightweight format with extremely high compression typically between 1/50 and 1/250.
3. With a single XVL file, measurements are also possible.
4. XVL has a good integration capability with MS Office documents, and integration with other systems is readily possible.
5. XVL has a small memory consumption.
6. XVL can be viewed on very inexpensive computers.
7. XVL has the capability of generating 2D illustrations.
8. XVL is accurate enough so that XVL data can be imported back to CAD systems.
9. XVL Player has good mouse operation integration on the freeware viewer.
10. XVL has optional encryption security.
11. XVL has constant 3D visualization quality in both zooming in or out.

MAN Nutzfahrzeuge thoroughly tested XVL, converting many test examples and ensuring the dependability and accuracy of the data as well as the scalability and security of the process. "For instance, the data of a bus frame that was 45 MB in CATIA can be compressed to 630 KB or less by XVL. And yet, the precision is maintained even if the data is vastly compressed. Anyone can obtain an exact dimension when they need to take a measurement of even the finest details. At MAN we have been very impressed with the high precision of XVL." (Dieter Ziethen)

They conducted tests on how data volume changes in a data conversion test flow such as CATIA V5 → .XV3 → IGES → STEP203 → CATIA V5. The results are shown in Figure 13.2. They found that XVL is by far the smallest format. While the IGES or STEP generated after this is very big, XVL was found to be light, but very accurate. Data such as structural information was retained, however they found some problems when data was returned back to CATIA, which they attributed to problems of the files generated from XVL. However, dimensions were accurate, posing no problems in practice. Convinced that XVL applications would help expand use of 3D data throughout the company and enable them to communicate, view, measure, and publish 3D data in a wide variety of ways, the company decided to adopt XVL.

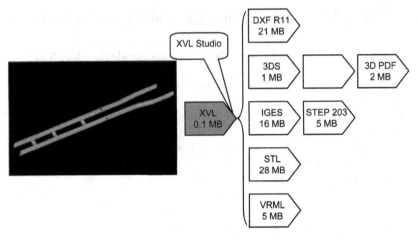

Figure 13.2 Results of XVL conversion test

12.4 Multi-use of XVL Centering Around Data Management Tools

Figure 13.3 shows the design process at MAN. It can be seen that XVL is effectively used in the downstream process. XVL is utilized as a communication means

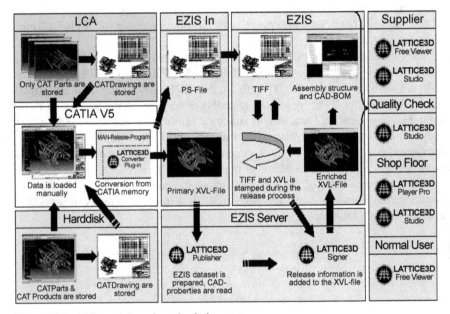

Figure 13.3 XVL used throughout the design process

in all processes; for example, in the design department for their discussion, information exchange with suppliers, and for communication between technical sales and their customers. To achieve this, the company established a system that ensures that staff requiring 3D data can access XVL easily.

Fortunately MAN had a company-wide product data management system (EZIS), in which they started to store XVL data. When a CAD designer checks in a CATIA model, an equivalent XVL file is automatically generated and stored in EZIS. This XVL is used as a product design viewing and discussion tool for use with suppliers and customers, *etc.* Specifically, the uses of XVL are as follows.

12.4.1 Internal Communication

One example is the coach department which recently introduced XVL and is currently in the process of storing coach data in EZIS. In EZIS, if a specific part is clicked from a parts list, various operations are possible including displaying a selected part view, deletion of part views, and automatic creation of an exploded parts views.

12.4.2 Communications with Suppliers

Instead of sending CATIA files which contain important design information, MAN will send XVL files or IGES files converted with XVL Studio to suppliers so that they can reference the MAN-defined geometry and other design information. For security purposes, instead of providing high accuracy data as it is, MAN uses XVL files with high but deliberately reduced accuracy.

12.4.3 Technical Illustrations

XVL serves as a very strong driving force for improving the efficiency of technical documentation because the use of 3D data enables development of technical documentation to be started at an earlier phase than before. XVL models can be animated and still retain their exceptional compression. Alternatively output can be an SVG illustration or an HTML page or 3D PDF. Furthermore, the company is planning to output 3D PDFs in Adobe format in the future. Making use of this flexibility of XVL, MAN plans to use 3D data in a broad range of areas for technical illustrations.

12.4.4 Assembly Instructions

Currently, process instructions and operational precautions are displayed using static images on a 32-inch large screen at the assembly lines in MAN's plants. MAN is however planning to display animated instructions to eliminate the time and labor required for preparing static images for visibly clearer and faster on-the-job guidance and training.

12.4.5 Quality Assurance

In terms of quality assurance, instead of using 2D paper-based views, MAN is planning to use 3D data for better confirmation of measuring points and geometry. Currently, special QA drawings are created from scratch each time for quality assurance but, in the future, 3D drawings, based on XVL and utilizing the existing 3D data, will be used to prepare such information, sharply reducing information exchange time; also useful information can be shared between different departments within the company and with customers.

According to Dieter Ziethen, "For the future we look forward to eliminating the time-consuming 2D drawing stage and only using more effective 3D digital models in as many areas as possible. We see many tasks becoming easier and simpler to understand as we move from static 2D drawings, crowded with information, to dynamic 3D models that deliver animated information at the time it is needed." The environment that MAN has built up of enabling lightweight XVL 3D data to be viewed on the more than 10,000 PCs at the company will no doubt serve as invincible competitive strength to the company.

Chapter 13
Using 3D Data Successfully

In the manufacturing industry, it is now common practice to implement 3D design using 3D CAD, review the 3D information by virtual simulation on computer (CAE), and transform this into a real object on a machine tool (CAM). But such uses of 3D data make up only a small part of the manufacturing process. Leading companies are starting to use 3D data not only in design, but also throughout manufacturing and even in downstream processes such as service and marketing. This chapter discusses the best practices in Japan, how to successfully use 3D data, and the key points for building a 3D data system.

13.1 Best Practices for Successful Use of 3D Data

The benefits of using 3D data are evident from the case studies that have been presented in this volume. 3D data itself has no value; value comes through use. Use of digital data replaces physical objects such as prototypes. With digital de-

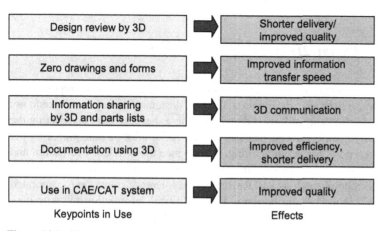

Keypoints in Use	Effects
Design review by 3D	Shorter delivery/ improved quality
Zero drawings and forms	Improved information transfer speed
Information sharing by 3D and parts lists	3D communication
Documentation using 3D	Improved efficiency, shorter delivery
Use in CAE/CAT system	Improved quality

Figure 14.1 3D uses in successful manufacturing businesses

velopment, physical "things" are not made until later in the process. This cost savings will continue to drive the expansion of digital processes. Within digital processes, the methods of using 3D data differ depending on the type and size of the business. But despite these differences, successful applications of 3D data share some common characteristics. The following are five successful patterns of using 3D data (Figure 14.1).

13.1.1 Design Review (DR) with Lightweight 3D

The first common use of lightweight 3D data is to enable DR between upstream and downstream departments. The importance of DR is well known; by pre-emptively resolving potential problems that may occur downstream, DR can shorten delivery time and increase quality. What to check during DR differs according to design stage and the nature of the business. DR using lightweight 3D data has two effects; the direct effects of reducing review time and work itself, and the indirect effects of preventing problems that may occur in the downstream process. The latter indirect effects are the very effects that prevent design changes, thus contributing to enormous cost reductions. In a concurrent design environment where the shape of the products undergoes routine changes every day, DR is vital for verifying the overall design. In addition, DRs using 3D data also have intangible effects. One is the interchange between experienced designers and newcomers through DR. Another is the fact that lightweight 3D can show the whole model, whereas CAD is limited to only subsets. This gives designers the overall view. Third is the fact that manufacturing staff and others who cannot read 2D drawings can now participate actively in DRs. Eliminating the 2D drawing "barrier," enables true collaboration between design and manufacturing. Since lightweight 3D data can be easily understood by manufacturing staff, it supports collaborative development between design and manufacturing. Bringing others with different perspectives into the development process drives innovation and can lead to revolutionary products.

13.1.2 Eliminating 2D Drawings and Reports

The second pattern is using lightweight 3D to eliminate drawings and reports from business processes. Even if design uses 3D data, design efficiency work would not improve if manufacturing still used drawings. This is because after completing the actual 3D design work, the designer has to then spend a lot of time preparing the drawings and reports for manufacturing. To enhance the efficiency of design and manufacturing, it is important to eliminate this process. This means that the information in the reports, such as surface finish and tolerance, must be incorporated into the 3D CAD model. Furthermore, that additional information must be converted into lightweight 3D data for downstream distribution. As discussed in previous chapters, some companies have succeeded in eliminating drawings and reports from their business processes. In order to achieve this, these companies have set standards for what is to be expressed in 3D, made rules for conveying information, and then established a system for automatically creating this information. The Japan Automobile Manufacturers' Association (JAMA) is currently trying to standardize the use of only 3D data for design information in the automobile industry. The goal is to replace 2D drawings by setting standards for 3D models and the viewers that interact with them. The elimination of 2D drawings from the process chain provides the following benefits: (1) it reduces the time and work for preparing drawings, (2) it makes it easier to understand 3D models, shapes, and product features, and (3) since no 2D drawings are prepared, there are no inconsistencies between drawings and 3D models. These 3D drawings will be used in many areas including production, services, and distribution. Therefore they must be expressed in lightweight 3D data format. No doubt, the use of 3D drawings will spread rapidly in the next few years.

13.1.3 Communicating with Lightweight 3D

The third common use of lightweight 3D data is for communication throughout the company and with partner companies. Information required by staff at manufacturing plants, quality control, and service departments is easier to understand in 3D rather than 2D drawings. Just being able to view 3D data on a notebook PC during discussions or while reading e-documents significantly improves understanding. A good example is the parts lists, where the use of 3D significantly reduces procurement errors. People from different backgrounds often have different levels of understanding. Use of 3D data helps such cases as it provides a common basic understanding of the model.

By sharing common understanding not only within the company but with parts suppliers, order placement and delivery errors can be prevented. Lightweight 3D viewers are available for free, so clients and partners can easily acquire them. Providing lightweight 3D drawings to partners and clients enables them to easily verify shapes. Lightweight 3D models increasingly contain dimensions, annotations, and other design intent, so subcontractors can now use them to manufacture parts. It is difficult to estimate the cost savings of using 3D data to communicate with clients and partners. However, like the Internet and e-mail, which are now indispensable for work, lightweight 3D data will also become essential once it is utilized in downstream processes.

13.1.4 3D Documentation

Another common pattern is using 3D data to generate illustrations for technical documents. The benefits of using 3D data include shorter delivery time, more consistent illustration quality, reduced illustration costs, and more. Beyond using

3D to create illustrations, it is also possible to add 3D data itself to documents. Parts catalogs require more than geometry, they also require configuration information. XVL contains such configuration information. It is also possible to create dynamic digital documents with 3D animation. For example, 3D data is useful for preparing assembly instructions for cell production locally and overseas. 3D interactive manuals are much easier to understand than paper manuals. Their scope is broad, ranging from manufacturing specifications to service and maintenance manuals. In the future, companies will even be able to distribute 3D instruction manuals to consumers.

The key to success lies in the acquisition of 3D data. Generally, the departments or subcontractors preparing manuals are far away from the design department and start work much later in the process. In order to implement 3D documentation it is necessary to establish a system that can quickly distribute 3D data to downstream departments. Access to 3D data by these departments can dramatically enhance their work efficiency. Data security can become an issue, especially when the manual creator is an outside company. Some manufacturers resolve the security by having specific 3D data non-disclosure agreements with their partners.

Lightweight 3D illustration and animation tools are still in the process of development, and will soon improve dramatically. Even so, tools are now available that can automatically update illustrations even when design changes occur. The potential cost savings are tremendous. The system for distributing 3D data need not be complex. As long as the rules for distributing 3D data are clearly understood, 3D data can be for manual creation right now. In the very near future, use of 3D data for illustrations will be a routine practice.

13.1.5 Sharing of 3D Data on CAE and CAT Systems

The fifth pattern of successful 3D data use lies the sharing of CAE and CAT data. Up to now, CAE and CAT data have only been used by a small group of specialists. But it has become apparent that sharing analysis data with the designers and manufacturers can enhance product quality. It is also useful to maintain and share past design failures with new engineers for educational purposes. And now, with lightweight 3D it is possible to reduce and share CAE data.

It is also advantageous to use 3D data at inspection departments. They are typically behind in adopting IT in their activities. Now with non-contact 3D measuring devices it is possible to detect differences between CAD models and the finished products, and share this difference data between different departments using lightweight 3D.

Such manufacturing process improvements require support from the manufacturing engineers, who are usually busy with their routine activities. Usually they are reluctant to go to the trouble of changing current business processes to take advantage of the 3D data. However, once they experience the convenience of 3D data, they usually become ardent supporters of 3D data. They then understand the advantages and they intuitively feel 3D will radically improve the manufacturing process. The applications of lightweight 3D data in sales, marketing, distribution, and service departments are steadily increasing. These successful uses of 3D data are building invisible competitive strength at the companies that are implementing them.

13.2 Systems that Aid in Successful Use of 3D Data

The next section discusses two key systems that are necessary to use 3D data successfully.

13.2.1 System for Storing 3D Data

Design information includes data before design approval and data after approval. It is the same for 3D data. Normally, DRs are carried out on 3D data before design approval, while factory parts checks are on 3D data after design approval. In either case, the key to success is to provide a system which allows access to the right 3D data whenever it is required. Without such a system, the search for data will limit work efficiency. DRs require pre-approved lightweight 3D data. All that is required for a system to deliver this data is to specify the location of the pre-approved CAD data and how to convert it. Once the location of this information is specified the company can then tell everyone the storage place of the data and the conversion rules. There are companies which distribute 3D parts lists, including pre-approved data, to plants to front-load work arrangements. There are also companies that share such data between with partners for DR purposes. One system that can provide pre-approval lightweight data to these limited users is a web-based automatic conversion system.

On the other hand, lightweight 3D data converted from post-design approval CAD data must be available to a broad range of users. To achieve this, the lightweight 3D data just needs to be stored in a database system that can be accessed by the whole company. For instance, post-design approval data is used for manual production and for quality assurance. In the case of design approval data, the CAD data and lightweight 3D data must be managed as a pair. Broadly speaking, such database systems can be implemented in two ways (Figure 14.2):

1. Expanding and using the file management system, or
2. Managing data in the PDM system.

The first system is inexpensive and can be implemented easily. It is suitable for small or medium-sized companies, but it does not have a database function in the real sense of the word. The second system is applicable to large manufacturers.

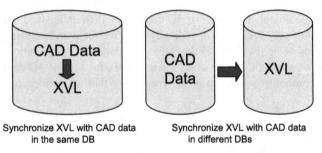

Synchronize XVL with CAD data   Synchronize XVL with CAD data
in the same DB   in different DBs

Figure 14.2   XVL storage mechanism

Most large manufacturers already have PDM, be it self-developed or a commercially available tool. So all that is needed is to integrate lightweight 3D into the existing system.

Sometimes the assembly structure and parts position information may differ depending on CAD systems. 3D data designed using different CAD systems should be corrected for consistency and saved in this data management system.

Often the lightweight 3D data will need to be modified when it is used in manuals. For example, parts catalogs often use names that are different from the part name defined in the CAD system. So it will usually be necessary to change part names and structure according to company rules. As long as the data is in XVL, there is software to fix this problem. The original XVL data converted from CAD can be transformed according to company rules. This allows the automatic generation of XVL data which matches company rules. The cost savings of this approach can be tremendous.

13.2.2 System for Ensuring Security

The greatest benefit of lightweight 3D data lies in the fact that 3D data can be displayed easily. This however can be a danger if the data leaks. Large amounts of confidential information can be compromised. Therefore is it imperative to build a reliable security system. However, restricting the data too much would make it difficult to use in downstream processes. So it is necessary to carefully balance the need for security with the need for convenience when establishing operational rules.

XVL technology has a function to control data access using passwords (Figure 14.3). It can be linked to the in-house security server allowing only personnel who have passed internal authorization to access XVL. In this case, internal rules on who can access what information when should be established before starting use. In many cases, there is a need to plan security taking into consideration not only the 3D data, but also relations with design documents, function charts, *etc*. So a more integrated security system is required. XVL technology allows such system integrations.

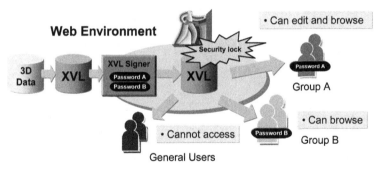

Figure 14.3   Security measures for using XVL

Some time ago, the Japanese manufacturing industry saw a boom in transferring manufacturing plants to China, but this has died down. Instead, Japanese manufacturers have invested in their local production facilities in attempts to promote technological innovation through the integration of development and production. The use of 3D data is effective for the integration of development and production systems in Japan, and also for design in Japan and production overseas, because the power of 3D communication and advantages of using 3D data for improving business processes are widely recognized.

But the story is not over, it is just beginning. There are still many ways to improve the manufacturing process using 3D data. For example, recently automobile manufacturers have started showing CAE analysis results to prospective customers at dealers to demonstrate the high quality of their cars. Companies are using 3D data in countless new areas: digital packaging systems using 3D data, sales and marketing systems using 3D contents, 3D instruction manuals for general consumers, and more. Generating lightweight 3D data and constructing a secure system for use opens the door to a broad range of data applications. Innovative use of lightweight 3D data will be a source of competitive strength for leading manufacturers for years to come.

Appendix A
Development Ideology

XVL (eXtensible Virtual world description Language) was developed to allow companies to make good use of what has become "idle" 3D data after design completes at their plants, maintenance divisions, marketing and sales divisions, *etc.*, in other words, throughout the whole company. Dramatically lightweight 3D data contributes to enhanced efficiency of the whole corporate process. At globally specialized manufacturing sites, employees with different skills and nationalities need to cooperate with each other in their work. Straightforward simple visual communication is indispensable in such environments. The following discusses the development concepts of XVL as a lightweight 3D format.

A.1 Lightweight:
The Starting Point of the XVL Development Concept

The idea of the lightweight of 3D data used in surface representation technology which forms the basis of XVL has been around since the 1980s. But back in those days, research on surface representation techniques aiming for use in CAD/CAM tended to focus on accuracy rather than size. It was in the mid 1990s that the authors began to refocus on lightweight representation methods using surface technology; this was about the time when the Internet started to spread. In 1997, VRML97 (Virtual Reality Modeling Language) was established as the ISO-standard 3D representation method in the Internet environment. The problem was, however, VRML97 expresses data using traditional aggregate clusters of polygons, and because telephone lines were mainly used in the Internet environment in those days, only very simple shapes could be sent. Though the concept of representing three dimensions in the network environment was innovative, it was not practical.

Lightweight surface technology can compress large CAD models by more than two orders of magnitude. The authors' idea was if data transfer time can be cut down by a factor of 100 then this would enable 3D to be built into the business processes of design and manufacturing, providing enormous benefits. This is how XVL research started out.

After three years of trial and error, lightweight XVL was born at Lattice Technology at the beginning of 2000. At this time, the development goals were:

1. To express 3D shapes accurately with only 1% of CAD data size. Ten times compression is not enough; at least 100 times compression is necessary for the data to be of use in business processes.
2. To convert all types of 3D data easily to XVL. Given that the costs to create 3D data are very high, it is not ideal to create XVL data from scratch. Be it CAD data, CG data, or costly scanned 3D data, a mechanism to automatically convert all 3D data is a must.
3. To allow the data to be web-enabled. The web is ubiquitous, so the data must be displayable in network environments.

By reducing CAD data, integrating it with production structural information, and also using net-friendly XML (eXtensible Markup Language), 3D XVL provides an ideal means of communication. XVL also supports comments and hyperlinks, turning 3D shapes into search tools.

Lattice also developed a technique to automatically generate data from diverse 3D data such as CAD and CG, enabling XVL to be generated from most 3D CAD/CG systems.

Based on the facts that XVL can be generated at low cost and that it can integrate design and manufacturing information with 3D shapes, XVL can be the reference for decisions in the design and manufacturing process. With XVL, 3D data can now be used easily throughout the design and manufacturing process. Applications include 3D design reviews, 3D parts lists, simple 3D email collaborations, 3D work instructions, and more. Because anyone can handle 3D data easily anywhere, Lattice calls this "3D Everywhere." It has been seven years since the birth of XVL and during this time the manufacturing industry has witnessed the spread of 3D CAD and the growing accumulation of 3D design data. The concept of "3D Everywhere" is increasingly being adopted by many companies to make further use of this 3D data.

A.2 From Fast Display to Information Unification

Figure A.1 shows five technical features of the initial XVL. Their details are discussed below.

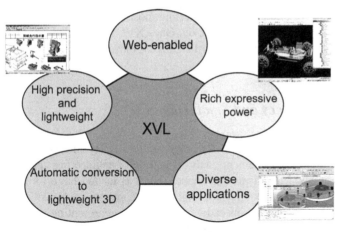

Figure A.1   Five characteristics of XVL

A.2.1   Automatic Generation of Lightweight 3D Data

The most important point to enable easy use of 3D data lies in the generation of inexpensive lightweight 3D data. Unlike image data, the generation of 3D data is expensive due to the large implementation, educational, and operational costs of 3D CAD. This is why it is crucial to be able to easily convert existing 3D CAD data to XVL.

XVL reduces the size of all 3D data, including 3D CAD, 3D CG data, and 3D scan data, providing a standard form for 3D expression. As shown in Figure A.2, XVL is about 100 times smaller than VRML, a polygon-based 3D format.

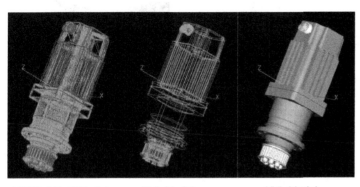

VRML 8 Split Model   XVL Model   XVL Model
3,201 Kbytes   32 Kbytes   Shading Drawing

Figure A.2   Comparison between VRML and XVL

At present, software for automatically converting most 3D CAD and CG data to XVL is commercially available on the market. Once XVL is generated, anyone can readily view the data with the freely distributed XVL Player.

A.2.2 High Data Accuracy and Small Data Size

Manufacturing departments use 3D data for cross-section display, measurement, NC tool path calculation, finite-element analysis, and other engineering applications. Polygon data is good for displaying shapes, but it is far too large for displaying models at the accuracy required for engineering applications. Since XVL uses surfaces and boundary curves instead of polygons for representing geometry, it can model CAD data to an accuracy of 1/1,000 mm with very small file sizes. When using XVL, only the lightweight curves and surfaces are transmitted over the network. The receiving PC uses this information to generate the image (Figure A.3). In addition, trimmed surface representations often used for complex shapes in CAD can also be saved in XVL format (a trimmed surface method is used to represent a complicated surface by using a normal square surface and boundary curves on the surface). The resultant P-XVL (Precise XVL) is thus both lightweight and high precision.

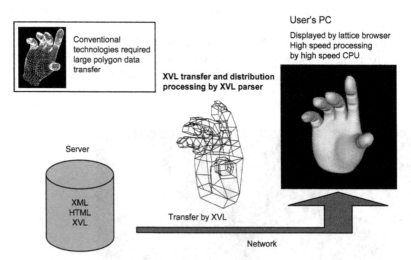

Figure A.3   High-speed 3D data transfer by XVL

A.2.3 Rich Expression

Another important feature of lightweight 3D data is the ability to add diverse information to the shapes. Lifelike images can be created by setting texture and color attributes. Mapping manufacturing data to the shapes presents manufacturing information in "visual form," and adding links from the shapes to other data enables "visual navigation."

XVL also provides a rich array of events and animation settings which display 3D processes clearly. XVL has been designed so that the animations can be controlled externally using languages such as JavaScript. It is also possible to launch 3D animations from external links, enabling fully interactive 3D manuals.

A.2.4 Network Compatibility

As mentioned above, XVL contains not only shape information but also product structure, dimensions, annotations, properties, materials, textures, and more. All of this XVL is based on XML (eXtensible Markup Language), which is a generalization of HTML, the "language of the web." This XML-based structure makes XVL easy to communicate over the web and easy to integrate with other structured documents. For example, the 2D SVG (Scalable Vector Graphics) format is also based on XML. Integrating SVG and XVL documents is easy, and makes it possible to navigate in 2D or 3D.

A.2.5 Diverse Applications

XVL enhances the efficiency of countless business processes. XVL applications range from engineering design reviews to 2D technical illustration production to generating quality assurance reports. XVL can be used alone, in interactive websites, in Microsoft Excel and Word documents and PowerPoint presentations, and in 3D PDF documents. And new applications can be written using XVL Kernel. The aim of XVL has been to reuse 3D CAD data, which requires enormous cost to generate, over and over again. Because XVL can combine shape information with other product data in a lightweight digital package, the applications are limitless. The basic concept of XVL is therefore to serve as the foundation for innovative 3D business processes within the company, with partners, and among general consumers.

A.3 New Upgraded XVL Technology Reduces Memory Consumption

With the growing use of 3D CAD data in the manufacturing industry, more and more companies are generating detailed 3D models. The data volume is increasing rapidly. For example, if all the parts of a car were to be expressed as precise 3D models, the total data volume would exceed 20 GB. It is impossible to display such models on standard PCs, where the memory size is usually limited to 2 GB. Sometimes models that are only 100 MB take several minutes to read and display, making them impractical for use.

A new version of XVL solves this problem. Through a new innovative surface technology which reduces memory consumption, XVL is able to display such models almost instantaneously. At Lattice, we call this XVL display technology V-XVL (Visual-XVL).

So XVL is not only lightweight and accurate, web-enabled, and has automatic conversion tools, it now also uses less memory and displays instantaneously. It can display CAD models exceeding 10 GB on PCs with only 2 GB of memory. It can display CAD models that even CAD systems cannot display. Figure A.4 shows a performance comparison of P-XVL and V-XVL. Both can compress a 300-MB SolidWorks file to about 3 MB. And V-XVL can display this model in 3 seconds using only 45 MB of memory.

With the growing use of 64-bit PCs, 64-bit-compatible CAD systems are also growing in use. Clearly, 3D models built in the design process will probably grow huger and huger, while the PCs used at plants, production management, and quality management departments remain 32-bit machines – sufficient for preparing documents and writing emails. The use of V-XVL technology resolves this situ-

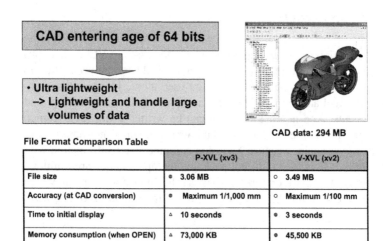

File Format Comparison Table

	P-XVL (xv3)	V-XVL (xv2)
File size	● 3.06 MB	○ 3.49 MB
Accuracy (at CAD conversion)	● Maximum 1/1,000 mm	○ Maximum 1/100 mm
Time to initial display	△ 10 seconds	● 3 seconds
Memory consumption (when OPEN)	△ 73,000 KB	● 45,500 KB

*Evaluation PC: Pentium-M 1.6 GHz Memory 1 GB

Figure A.4 Comparison between P-XVL and V-XVL

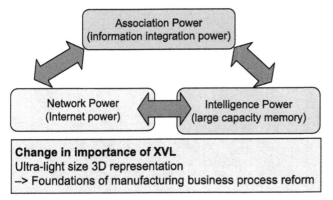

Figure A.5   Change in XVL position

ation as it allows inexpensive PCs to display gigantic 3D data. Even the free
XVL Players can display these enormous models. This means that, by converting
the data to XVL, massive 3D models designed on 64-bit PCs at leading companies
can be used by small subcontractors who only have 32-bit PCs.

Figure A.5 shows the main features of today's XVL, namely, information integration ability using XML, small size which promotes data sharing on the network, and efficient memory use which enables the display of massive data. In the
seven years since its birth, XVL has transitioned from a tool for ultra-lightweight
3D display to a "foundation for manufacturing business process innovation." Increasing numbers of manufacturers are generating, storing, and sharing 3D XVL
data as they improve their business processes.

Appendix B
Overview of XVL Products

This appendix introduces the main XVL applications that are available on the market. The easiest way to use XVL is to install the free "XVL Player" and start viewing models. You can download XVL Player from www.lattice3d.com, then see the demo page to view XVL models. Companies using XVL for communication often have XVL Player installed on every computer.

Basically, XVL applications can be categorized into three groups – conversion, editing, and publishing (Figure B.1). Conversion software converts data from CAD to XVL (Table B.1). Editing software modifies the XVL data, for example by adding annotations and comments, changing structural information, and creating animations. Table B.2 summarizes the functions of the "XVL Studio"

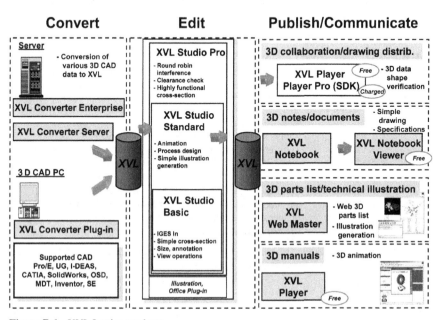

Figure B.1 XVL Lattice products

family of XVL editors. Publishing software creates interactive 3D documents based on XVL and writes them to different formats, including HTML, MS Office, and 3D PDF (Table B.3). Together, these tools enable 3D data to be used for parts lists, assembly instructions, quality control, design reviews, and other shape-based activities. The following sections introduce XVL tools from these three function groups.

Table B.1 Main XVL conversion product groups

Purpose	Software	Functions
Conversion	XVL Converter Series 1. XVL Converter Plug-in 2. XVL Converter Light	Converts CAD or CG data to XVL 1. Plugs into the CAD system and provides a "save as XVL" function 2. Standalone converter for CAD data. Supports batch and automatic conversion Available from Lattice Technology and CAD developers CAD formats include CATIA V4, CATIA V5, UG, SolidEdge, Pro/Engineer, Solid Works, Inventor, One Space Designer, and others
Security	XVL Signer (+ public key encryption option)	Signs and encrypts XVL files, so that only authorized personnel may use them. Can work alone or with an authentication server
Process	XVL System Toolkit	Set of command-line tools that process XVL files. Used to develop automated XVL processing systems

Table B.2 XVL Studio product family

Purpose	Software	Function
Viewing	XVL Studio Basic	Displays XVL geometry, generates cross-sections, measures dimensions. Simple editing of assembly structure and annotations, *etc.*
Editing	XVL Studio Standard	Tool for supporting technical documentation. Has functions to create 3D animations and process plans. Includes XVL Studio Basic functions
DR	XVL Studio Pro	Provides design review functions such as interference check, advanced cross-section generation, interference reporting. Includes XVL Studio Standard functions
Illustration	Illustration options	Creates illustrations from XVL models

Table B.3 3D documentation product groups

Purpose	Software	Functions
3D documents	XVL Notebook Standard	3D document editor. Creates interactive 3D documents that include 3D models, 2D illustrations, assembly trees, product configuration information, markups, and more. Outputs to XVL, MS Office, HTML, and 3D PDF
Interactive 3D web pages	XVL Web Master	Automatically generates interactive 3D web pages such as 3D parts lists and 3D manuals

B.1 XVL Studio Basic: Viewing and Editing Models

Production uses of 3D data include shape measurement and cross-section display. XVL Studio Basic serves these purposes by providing basic viewing and editing functions. As shown in Figure B.2, this tool has functions for editing annotations, changing colors, transforming parts, and editing the assembly structure. It is able to read and write several 3D formats including XVL and IGES.

It is increasingly common for subcontractors to receive CAD models – not drawings – from their customers. Suppliers often receive CAD data in different formats from different customers. For these suppliers it is not economical to install

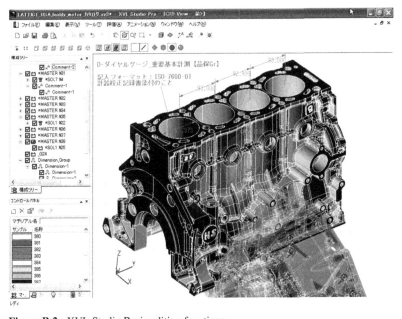

Figure B.2 XVL Studio Basic editing functions

multiple CAD systems just to view the different CAD models. XVL Studio Basic is the perfect solution for such suppliers.

At large companies, XVL Studio Basic serves as a useful and economical tool for 3D communication, particularly in areas that do not use 3D CAD such as manufacturing plants and production technology departments. Efficient use of XVL Studio Basic eliminates the need for drawings. Traditionally drawings have been used to convey shape, dimensions, and instructions to downstream departments. XVL Studio Basic can display the same information in XVL models.

In this way, XVL eliminates the need for analog processes such as drawings, enabling design information to be digitally sent to downstream processes. Providing such production information to business partners further reinforces collaborative business relationships. Such tight partner relationships may become key sources of competitive strength.

B.2 Products for Document Creation Using XVL

XVL is commonly used to create technical documents. There are two basic methods:

1. Attaching 3D models to e-documents and
2. Creating 2D images or illustrations from 3D models, and pasting these into e-documents

3D models are widely used for parts catalogs and assembly manuals. 3D assembly animations, for example, are clear and easy to understand. However, paper technical documents are still widely used. 3D data can streamline the process of creating paper documents by automatically generating the 2D images and illustrations.

B.2.1 XVL Studio Standard: Creating Animations

XVL Studio Standard provides several methods of creating animations (Figure B.3). It includes a process planning feature, which enables the user to create an animation just by specifying the order in which the parts come together. All the user needs to do is to define the tree describing the assembly order. This tool automatically generates animation from this definition. The assembly animation can be played back on XVL Player, the free XVL viewing tool.

Another animation method in XVL Studio Standard is keyframe animation. XVL Studio Standard provides the standard keyframe editing tools that are familiar to any experienced animator.

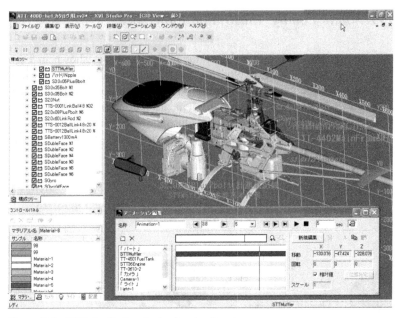

Figure B.3   XVL Studio Standard animation functions

XVL Studio Standard also has a function for defining snapshots. As shown in Figure B.4, snapshots can contain camera positions and part positions, visibility, and rendering style. Sometimes, 3D data can be too rich and complex. Snapshots are a convenient way to reduce confusion and highlight particular aspects or features of a model, thus improving communication.

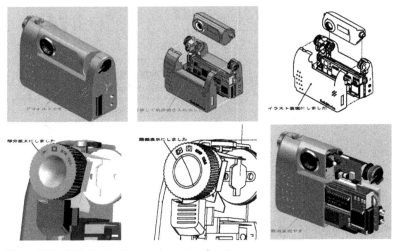

Figure B.4   XVL Studio Standard snapshot functions

B.2.2 XVL Studio Family: Illustration Functions

XVL Studio also comes with optional illustration functions. As shown in Figure B.5, XVL Studio can generate exploded views for parts catalogs. XVL Studio enables the user to define the part disassembly order, part trajectories, rendering styles, part numbers, and other illustration elements. Furthermore, all of these elements can be easily edited. The resulting illustrations can be saved in raster and vector formats. They can then be edited in illustration programs or inserted directly in documents. Use of 3D data dramatically reduces illustration time.

Along with part geometry, XVL also includes part attributes. XVL Studio can output part attribute information in CSV format. The part attribute information can be combined with illustrations to create the parts catalogs shown in Figure B.6.

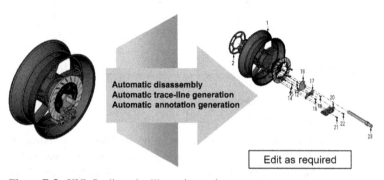

Automatic disassembly
Automatic trace-line generation
Automatic annotation generation

Edit as required

Figure B.5 XVL Studio series illustration option

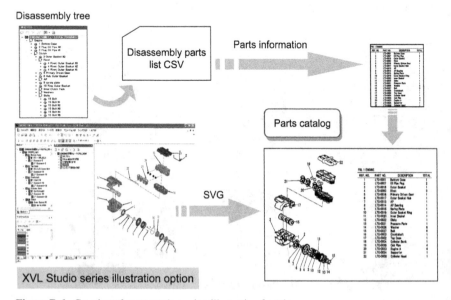

Disassembly tree

Disassembly parts
list CSV

Parts information

Parts catalog

SVG

XVL Studio series illustration option

Figure B.6 Creation of parts catalog using illustration function

B.2.3 XVL Notebook: 3D Document Creation

XVL Notebook enables the user to easily create 3D digital documents. As shown in Figure B.7, XVL Notebook is able to display 3D shapes, images, tables, and text information all together. Just pasting an XVL file into a Notebook document will create a 3D view. The 3D view is fully interactive; it enables the user to pan, zoom, and rotate the model. The 3D view image can be saved as a snapshot. Also,

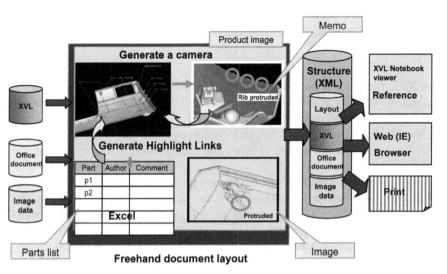

Figure B.7   XVL Notebook document editing

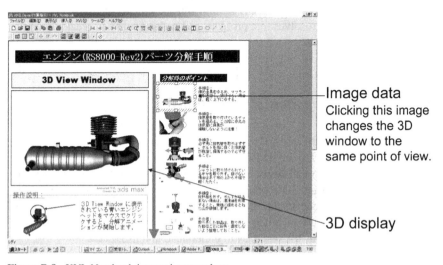

Image data
Clicking this image changes the 3D window to the same point of view.

3D display

Figure B.8   XVL Notebook interactive snapshots

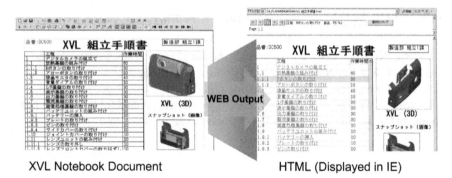

XVL Notebook Document   HTML (Displayed in IE)

Figure B.9   XVL Notebook Standard web page generation for assembly process instruction

the XVL file has assembly information and part data. This information can be extracted and saved as tables in the document. Furthermore, all these elements, 3D views, 2D snapshots, and table, are linked together. For example, selecting a part in the 3D view can highlight that part's information in the table. And selecting a 2D snapshot will reorient the 3D view and change the rendering style so that it matches the snapshot (Figure B.8).

XVL Notebook creates dramatically new kinds of documents. While paper documents are stationary in nature and do not change, Notebook documents change dynamically, allowing the user to view the data from any desired viewpoint. Notebook documents can also play 3D animations defined by XVL Studio Standard, making Notebook a fast and easy tool for creating work procedures and disassembly manuals. Notebook documents can be saved as XVL, 3D PDF, and HTML (Figure B.9), enabling them to be shared over the internet.

In the past it has not been difficult to create documents with 3D views, but it has been very difficult to link the 3D data to other elements such as images and tables. XVL Notebook resolves that problem and makes it easy to create dynamic, interactive 3D documents.

B.2.4 Lattice3D Reporter: 3D Spreadsheets

Many companies use Microsoft Excel to generate reports, check sheets, and other production documents. Lattice3D Reporter is an Excel add-in that enables the user to insert 3D data into spreadsheets. Similar to XVL Notebook, Lattice3D Reporter enables the user to insert 3D models, 2D images, part tables, buttons, and more into spreadsheets. Of course the elements are linked, so selecting a part in one view highlights that part in other views. This turns ordinary spreadsheets into interactive 3D documents that are easy to understand and use.

B.2.5 XVL Web Master: Automatic Webpage Generation

XVL Notebook is great for manually producing interactive 3D documents, but as companies expand their use of 3D they need a way to automatically generate such documents. XVL Web Master fills that need. It automatically processes XVL files to produce parts lists, assembly instructions, and other interactive 3D HTML pages (Figure B.10). These pages can be served on the web, and clients only need a web browser to view them from anywhere in the world.

XVL Web Master contains several advanced functions. Given an XVL file it can automatically create 2D images and illustrations. These illustrations can be linked to the 3D data, so that when the user selects a part in the illustration it highlights in the 3D model, and vice versa. XVL Web Master can also add disassembly animations which enables the user to "explode" the model in the 3D view.

But the most valuable function of XVL Web Master is its ability to integrate product information with the 3D model. XVL files contain structure information and part data along with the 3D shapes. XVL Web Master can extract this information and put it in assembly trees and parts tables on the page. Of course these trees and tables are linked to the 3D and 2D views, so that selecting a part in one view will highlight it in the others. But XVL Web Master goes beyond just extracting the data that is already in the XVL file; it can also integrate external information into the page. For example, most companies keep part data in a database separate from their CAD models. XVL Web Master can automatically add such external part data to the part table and make it available to all users.

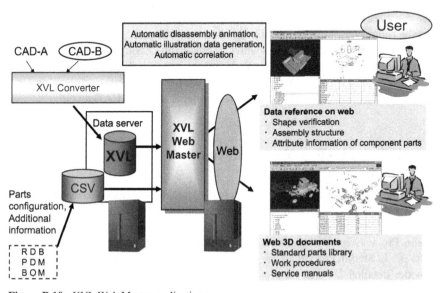

Figure B.10   XVL Web Master applications

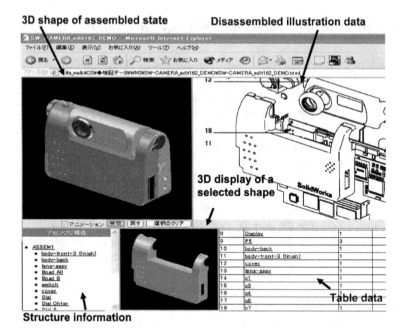

Figure B.11 XVL Web Master example output

Figure B.11 shows a webpage generated by XVL Web Master. Four types of information are displayed here: a 3D view (XVL) on the top left and bottom center, a 2D illustration (SVG) on the top right, an assembly tree on the bottom left, and a parts table on the bottom right. Clicking a part in any of the views highlights that part in the other views, so, for example, by selecting a part in the 3D view the user can check its part number, supplier, and price at a glance. The table could even be further customized to enable them to order the part.

XVL Web Master supports batch and automatic processing of XVL files. It provides an economical to way to generate large numbers of interactive 3D documents. Many companies use XVL Web Master to automatically generate parts lists, assembly instructions, and other interactive 3D documents for every CAD model.

B.3 XVL Studio Pro: Design Review

In the manufacturing industry, design changes incur the greatest cost when they happen late in the production phase. Design reviews (DRs) are conducted during the design phase to uncover problems and prevent late-stage design changes. XVL Studio Pro was jointly developed by TOYOTA and Lattice to enable efficient DRs. XVL Studio Pro calculates interferences and clearances on large models, provides detailed 2D and 3D cross-sections, generates illustrations of each interference, and writes DR reports.

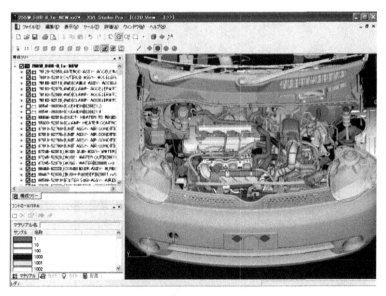

Figure B.12   XVL Studio Pro large model display

As shown in Figure B.12, XVL Studio Pro handles models containing thousands of parts, meaning that CAD data exceeding 10 GB can be displayed on PCs. XVL Studio Pro can also simultaneously display 2D and 3D cross-sections (Figure B.13). This feature allows users to interactively section even very large models, and the 2D and 3D views will update dynamically. This is useful for DRs,

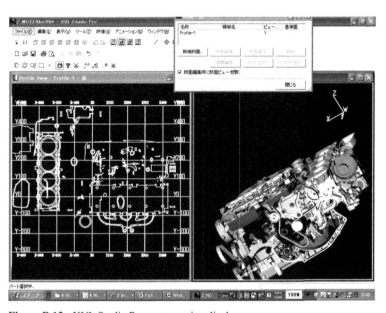

Figure B.13   XVL Studio Pro cross-section display

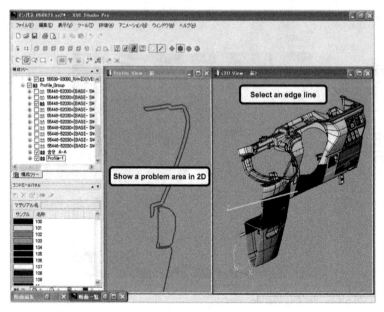

Figure B.14   Cross-section from edge line

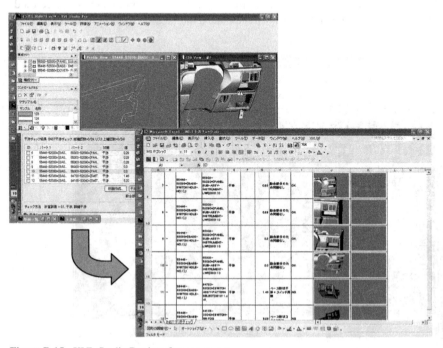

Figure B.15   XVL Studio Pro interference report

during which interferences, clearances, and contacts can be observed in detail. As shown in Figure B.14, XVL Studio Pro can also extract cross-sections along curves, making it useful to designers of complex products. And finally, XVL Studio Pro can generate interference reports (Figure B.15). These reports not only list the interferences; they also include illustrations of the interference areas. These illustrations show the problems at a glance. The reports themselves enable problems to be easily tracked, managed, and resolved.

XVL Studio Pro increases the efficiency and value of DRs. It enables design reviews of whole models, not just selected pieces. Automatic interference checking is fast and complete; there is no need to check again after the first time. Interference reports highlight the problems, so the meeting can focus on resolving them. Dynamic 2D and 3D cross-sections enable reviewers to examine problems in detail. And the use of 3D data enables non-engineers such as plant staff to participate in reviews. Such personnel might have problems interpreting the design from just drawings. In this way, XVL Studio Pro lowers the barriers between design and manufacturing and contributes to smooth communication between the two parties. This, in turn, improves design quality.

B.4 XVL Signer: Security Tool

XVL data is lightweight and easy to use. It can be sent by email and viewed by anyone using the free XVL Player. These benefits, however, come with some risk. XVL, being lightweight, has the potential to leak and spread important design information outside of the organization. To prevent this, XVL data can be secured using XVL Signer. XVL Signer is a tool which signs and encrypts XVL files, allowing only those who know this password to access the data (Figure B.16).

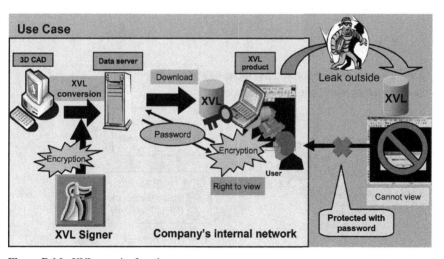

Figure B.16 XVL security function

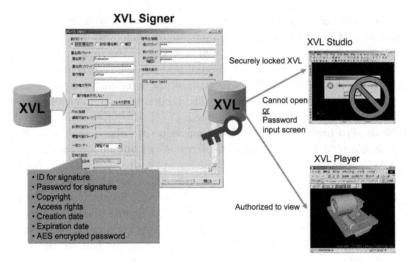

Figure B.17   XVL Signer function

XVL Signer adds four levels of access rights to XVL files: editing rights, measurement rights, browsing rights, and access rights. Some users may be able to open the file but not measure it; others may not be able to open the file at all. As shown in Figure B.17, XVL Signer can also set the creation and expiration date of an XVL file. When the term expires, the file will immediately be rendered inaccessible, and none will be able to view it. In addition, XVL Signer can set the copyright string that will be displayed with the model in XVL Player.

Leading companies often have security servers with user authentication functions to control access to in-house information. XVL Signer is designed to integrate with such authentication systems and provide seamless protection for XVL files and related data (Figure B.18).

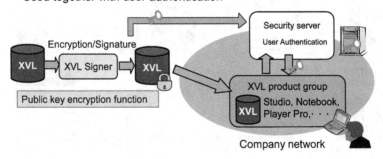

Figure B.18   Public key encryption function

Postscript

Innovation comes in two ways: reactive and proactive. The use of 3D CAD data in product design can be said to be reactive innovation and defensive reform to catch up in competition and business partnerships with rival manufacturers and leading manufacturers overseas. The use of 3D data in the areas of manufacturing and maintenance has indeed enabled Japan to assume a leadership role in the world. 3D data is a tool which allows pioneering proactive innovation and manufacturing reforms.

Japan's manufacturing industry is known for its high-quality products based on its long tradition of design work carried out jointly by designers and production staff. This practice has enabled the two parties to talk about and resolve problems encountered at the manufacturing site, and is the greatest reason why Japanese products have been able to preserve their quality during the high-growth period of the Japanese economy.

However, today, Japanese firms are transferring not only their production bases, but also their development bases overseas, creating a situation where they are finding communication becoming increasingly difficult. For a while it seemed that the country was losing its competitive strength due to this increasing loss of communication, but actually this trend has brought about the rapid 3D manufacturing process innovations described in this book. There are even companies that are innovating manufacturing process based on 3D data throughout the organization, from their design departments to manufacturing, procurement, quality assurance, marketing, and maintenance departments. What these companies are aiming at using IT for is the reinforcement of cooperation and dialog internally between their different departments.

Process innovation methods are diverse, and depend on the business type and operations. There is neither one universal method nor business solution. In this sense, 3D manufacturing is clearly a proactive reform. With more and more companies accumulating 3D data, these companies are now able to sustain their acquired competitive strength over a long time just by improving the methods of using 3D data. The application of IT tools for accumulated knowledge overflowing from the manufacturing site will increasingly strengthen the Japanese manufacturing industry.

The twenty-first century is said to be an era for those in control of knowledge and information. In rapidly changing times, the speed at which knowledge becomes obsolete will accelerate even more. This will be accompanied by the need to accumulate knowledge in the company, convey it, and share it even more quickly. Needless to say, intellectual labor cannot be improved just by pep talks. Rules for supporting sound stable intellectual labor are required, the most reliable of which is the framework for using 3D data. Today, as manufacturers arm themselves with IT as a tool to win in the global competition, it is hoped that this book will contribute in some way by providing beneficial information.

Index

2

2D drawing interchange file (DXF) 63

3

32-bit machine 134
32-bit PC 9
3D data vi
3D data information distribution platform 25
3D document 145
  3D catalog 111
  3D digital document 143
  3D instruction manual 124
  3D interactive manual 124
  3D manual 42
  3D parts list 38
  3D visual manual 43
  interactive 3D document 144
  technical 3D document 111
3D drawing 120, 123
3D Everywhere 13, 130
3D manufacturing 151
3D PDF 8, 133, 138
3D viewer 79, 83, 91
3D XML 7

6

64-bit PC 9, 134

A

ALPINE PRECISION 34, 81
animation 140
  3D process animation 34, 110
assembly instruction 69, 120

assembly procedure 110
automating the translation and flow of information 78

B

Bill Of Materials (BOM) 28
broadband internet 2

C

CAD 27
  3D CAD vi, 2
  CATIA 7, 117
  Pro/ENGINEER 99
  SolidWorks 110
  Unigraphic 71
CAE 124
CASIO 34, 99
CAT 124
clearance check 37
collaborative manufacturing 34
combination manufacturing 34
comma separated value (CSV) 38
communication pipeline 63
complete digitalization 69
computer-aided engineering (CAE) 46
computer-aided testing (CAT) 48
contour map 47
core of design 57
CRIC cycle ix
cross-section 36, 53

D

data transfer time 100
design and manufacturing process innovation 60

design information v, 4, 23
design intent 57, 72, 83, 110
design quality 58
design review (DR) vii, 33, 34, 51, 60, 84,
    97, 122, 146
    online design review (ODR) 107
documentation 14
downstream 13
downstream process 63
drawing 15
drawing-less vi, 85, 88, 96
DWF 8

E

ECMA 8
eDrawing 110
electric discharge machining (EDM) 92
eliminating drawing and report 123
e-manual 101
Engineering Change Order (ECO) 112
ERP system 114

F

file management system 125
frontload 101
    front-loading 67

H

HTML (Hyper Text Markup Language)
    12

I

IGES 27, 139
illustration 142
Information Technology (IT) 1
instruction manual 34, 102
interference 36, 53
interference checking 58
interference problem 36
interference report 149

J

Japan Automobile Manufacturers'
    Association (JAMA) 123
JT 8

K

Koutei 75
KVAL 34, 109

L

L-3 COMMUNICATIONS (L-3C)
    34, 109
lattice technology vi, ix
lightweight 3D 10
lightweight 3D data vi, 11, 131
liquid crystal display (LCD) 59

M

MAN 34
MAN Nutzfahrzeuge AG 115
manufacturing information 17
master data 17
Microsoft Excel 144
MONOZUKURI v, 7

N

NC programming 93
NIKON 33, 59

O

official data 17
original drawing data 27
owner manual 101

P

paperless communication 84
parts catalog 34
PDF 27
PDM 125
polygon 8
process management system 73, 74
product assembly 34
product data management (PDM) 3
product data quality (PDQ) 4
product development cycle 99
product lifecycle 91
product lifecycle management (PLM) vii
product maintenance 34
production instruction 69
prototype review 28

Q

quality assurance 120

R

report 34, 82, 144
report-less 88
ROI 115

S

security 67, 89, 97, 126, 150
semiconductor 59
service manual 101
sheet metal DR 55
simultaneous engineering 51
Single Source Multi Use 99
snapshot 141
SONY 21
SVG (Scalable Vector Graphics) 133

T

technical document 123
technical documentation 119
technological barrier 110
TOKAI RIKA 34, 91
total cost of ownership (TCO) 22
TOYOTA 33
TOYOTA MOTOR CORPORATION
    (TOYOTA) 51

U

U3D 8

V

very large data 14
viewer 12

VRML 8
VRML97 (Virtual Reality Modeling
    Language) 129

X

XML (eXtensible Markup Language) 12,
    130, 133
XVL vi, ix, 9, 27, 129
    P-XVL (Precise XVL) 132
    V-XVL 67
    V-XVL (Visual XVL) 134
    XV2 67
    XVL Kernel 133
    XVL-based communication system
        87
XVL product 137
    Lattice3D Reporter 144
    XVL Notebook 71, 93, 143
    XVL Player 115, 132, 137
    XVL Signer 149
    XVL Studio 85, 101, 137
    XVL Studio Basic 139
    XVL Studio Pro 52, 146
    XVL Studio Standard 140
    XVL Web Master 67, 71, 85, 97,
        112, 145

Y

YAMAGATA CASIO 34, 69